T0137245

UNIPA Springer Series

The **UNIPA Springer Series** publishes single and co-authored thematic collected volumes, monographs, handbooks and advanced textbooks, conference proceedings, professional books, SpringerBriefs, journals on specific issues of particular relevance in six core scientific areas. The issues may be interdisciplinary or within one specific area of interest. Manuscripts are invited for publication in the following fields of study:

1- Clinical Medicine;
2- Biomedical and Life Sciences;
3- Engineering and Physical Sciences;
4- Mathematics, Statistics and Computer Science;
5- Business, Economics and Law;
6- Human, Behavioral and Social Sciences.

Manuscripts submitted to the series are peer reviewed for scientific rigor followed by the usual Springer standards of editing, production, marketing and distribution. The series will allow authors to showcase their research within the context of a dynamic multidisciplinary platform. The series is open to academics from the University of Palermo but also from other universities around the world. Both scientific and teaching contributions are welcome in this series. The editorial products are addressed to researchers and students and will be published in the English language.

More information about this series at http://www.springer.com/series/13175

Francesco Lo Piccolo · Marco Picone · Vincenzo Todaro

Editors

Urban Regionalisation Processes

Governance of Post-Urban Phenomena in Sicily

UNIVERSITÀ
DEGLI STUDI
DI PALERMO

Springer

Editors
Francesco Lo Piccolo
DARCH
University of Palermo
Palermo, Italy

Marco Picone
DARCH
University of Palermo
Palermo, Italy

Vincenzo Todaro
DARCH
University of Palermo
Palermo, Italy

ISSN 2366-7516   ISSN 2366-7524   (electronic)
UNIPA Springer Series
ISBN 978-3-030-64471-0   ISBN 978-3-030-64469-7   (eBook)
https://doi.org/10.1007/978-3-030-64469-7

This Springer imprint is published by the registered company Springer Nature Switzerland AG
The registered company address is: Gewerbestrasse 11, 6330 Cham, Switzerland

Acknowledgments

This book presents the results of the research activities of the Research Unit of the University of Palermo (local coordinator Professor Francesco Lo Piccolo) of the National Interest Research Project (PRIN 2010–2011) "Post-metropolitan territories as emergent forms of urban space: coping with sustainability, habitability, and governance" (*Territori post-metropolitani come forme urbane emergenti: le sfide della sostenibilità, abitabilità e governabilità*) funded by the Italian Ministry for Education, University and Research (MIUR)—national coordinator Professor Alessandro Balducci (Politecnico di Milano).

Contents

Institutional Policies

Abbreviations

APAT	*Agenzia per la Protezione dell'Ambiente e per i Servizi Tecnici* (Agency for Environmental Protection and Technical Services)
ARS	*Assemblea Regionale Siciliana* (Sicilian Regional Assembly)
ATA	*Alta Tensione Abitativa* (High Housing Tension)
CBA	Commonwealth Bank of Australia
CEP	*Comitato di Coordinamento dell'Edilizia Popolare* (Coordination Committee for Public Housing)
CIPE	*Comitato Interministeriale per la Programmazione Economica* (Inter-Ministerial Committee for Economic Programming)
CIRCES	*Centro Interdipartimentale sui Centri Storici* (Interdepartmental Research Centre on Historical Towns)
DdL	*Disegno di Legge* (Draft Law)
DI	*Decreto Inter-Ministeriale* (Inter-Ministerial Decree)
DL	*Decreto Legge* (Law Decree)
Dlgs	*Decreto legislativo* (Legislative Decree)
DM	*Decreto Ministeriale* (Ministerial Decree)
DOC	*Denominazione di Origine Controllata* (Controlled and Guaranteed Designation of Origin, CGDO)
DPR	*Decreto del Presidente della Repubblica* (National Presidential Decree)
DR	*Decreto del Presidente della Regione* (Regional Presidential Decree)
EIA	Environmental Impact Assessment
EIncA	Environmental Incidence Assessment
ERP	*Edilizia Residenziale Pubblica* (Residential Public Building)
FIAT	*Fabbrica Italiana Automobili Torino*
GAL	*Gruppo di Azione Locale* (Local Action Group)
GDP	Gross Domestic Product
HIA	Housing Industry Association
IACP	*Istituto Autonomo Case Popolari* (Autonomous Council Housing Institute)
IDA	Industrial Development Areas

IDFP	Index of Dispersion of the Foreign Population
IGP	*Indicazione Geografica Protetta* (Protected Geographical Indication, PGI-quality)
ISPRA	*Istituto Superiore per la Protezione e la Ricerca Ambientale* (Institute for Environmental Protection and Research)
ISTAT	*Istituto Nazionale di Statistica* (National Institute of Statistics)
ITEC	Information Technology, Electronics and Communications
MA	Metropolitan Area
MC	Metropolitan City
MEF	*Ministero dell'Economia e delle Finanze* (Ministry of Economy and Finance)
NAR	National Association Realtors
NL	National Law
OMI	*Osservatorio del Mercato Immobiliare* (Observatory of Real Estate Market)
ONPI	*Opera Nazionale per i Pensionati d'Italia* (National Body for Italian Pensioners)
OPCER	*Opera Pia Cardinale Ernesto Ruffini*
PEEP	*Piano per l'Edilizia Economica e Popolare* (Plan for Popular and Low-Cost Construction)
PII	*Programma Integrato di intervento* (Integrated Intervention Programme)
PLANECO	Planning in Ecological Network
PON METRO	*Programma Operativo Nazionale Città Metropolitane 2014–2020* (National Operational Programme for Metropolitan Cities 2014–2020)
PRG	*Piano Regolatore Generale* (Master Plan)
PRIU	*Programma di Riqualificazione Urbana* (Urban Renewal Programme)
PRU	*Programma di Recupero Urbano* (Urban Recovery Programme)
PRUSST	*Programma di Riqualificazione Urbana e Sviluppo Sostenibile del Territorio* (Urban Renewal and Territorial Sustainable Development Programme)
PTPR	*Piano Territoriale Paesistico Regionale* (Landscape Regional Plan)
QCS	*Quadro Conoscitivo con valenza Strutturale* (Regional Framework with Structural Value)
QPS	*Quadro Propositivo con Valenza Strategica* (Proactive Framework with Strategic Value)
RAI	*Radio Televisione Italiana* (National Italian Television)
RL	Regional Law
RNO	*Riserva Naturale Orientata* (Oriented Nature Reserve)
SAC	Special Area of Conservation
SCI	Site of Community Importance
SEA	Strategic Environmental Assessment

SICET	*Sindacato Italiano Casa e Territorio* (Housing and Territory Italian Syndicate)
SOSVIMA	*Agenzia di Sviluppo locale delle Madonie* (Madonie Local Development Agency)
SPA	Special Protection Area
SUNIA	*Sindacato Unitario Nazionale Inquilini e Assegnatari*
TAR	*Tribunale Amministrativo Regionale* (Regional Administrative Court)
UAA	Utilised Agricultural Area
UNESCO	United Nations Educational, Scientific and Cultural Organisation
WHL	World Heritage List
ZEN	*Zona Espansione Nord* (Northern Development Area)

Urban Regionalisation Processes in Sicily: From the Theoretical Framework to the Local Dynamics

Francesco Lo Piccolo, Annalisa Giampino, Marco Picone, and Vincenzo Todaro

Abstract This introductory chapter proposes a mainly theoretical reflection on the category of the urban, both as a spatial and social product and as an economic and political one, in territories marked by settlement transformation phenomena. In tracking this evolutionary path, this chapter addresses the huge thematic bibliography on the phenomena of urban regionalisation without any demand for systematic approach but pursuing the aim of tracing the *geography* of this on-going transition in Sicily.

1 For a Critical Theory on Urban Regionalisation Processes

The critical reflection of modern cities, that continued through the twentieth century, has fed a wide-ranging and controversial debate regarding urban studies. The far-reaching modification processes of economic, political and social assets, associated with spatial reorganisation and restructuring processes, have marked a radical *break away* from traditional transformation tendencies (Dear and Flusty 1998), epistemologically creating the crisis of interpretive binary and dichotomous paradigms at the basis of the theoretical meaning that—in the *city* and, in its *ontological opposite*, rural areas—traced the understandability of the field of urban studies (Brenner 2016).

F. Lo Piccolo (✉) · A. Giampino · M. Picone · V. Todaro
University of Palermo, Palermo, Italy
e-mail: francesco.lopiccolo@unipa.it

A. Giampino
e-mail: annalisa.giampino@unipa.it

M. Picone
e-mail: marco.picone@unipa.it

V. Todaro
e-mail: vincenzo.todaro@unipa.it

© The Editor(s) (if applicable) and The Author(s), under exclusive license to Springer Nature Switzerland AG 2021
F. Lo Piccolo et al. (eds.), *Urban Regionalisation Processes*, UNIPA Springer Series,
https://doi.org/10.1007/978-3-030-64469-7_1

It is this very theoretical category of *urban* both as a spatial, social product and as an economic and political one, that has changed via new settlement principles, an expression of *new forms of cities* and also of new *city demand* (Balducci and Fedeli 2007) and of the various use practices that produce and are produced in these new territories. This pluralisation of the urban location can be traced to the semantic hordes of neologisms produced by urban research in a rather short space of time which, from the second half of the twentieth century to date, shows us the increasingly strategic role played by the urban phenomenon and the "blind field", as defined by Lefebvre (1970), in which urban studies seem to be bridled. A process of empirical knowledge accumulation on the important aspects of urban space, without an essential conceptualisation of the "new theoretical geographies of urban spaces" (Roy 2009, 820).

Literature on this topic (including, in this vague yet broad definition, the heterogeneous range or research and realms that can be attributed to *urban studies*) has contended with a large number of investigation and classification approaches and modes, with the aim of describing the changes in urbanisation phenomena. The result is the production of a number of studies and consequent definitions: multicentric metropolis, post-metropolis, city-region, multicentric city, mega-city, planetary urbanisation, regional urbanisation, infinite city, and all the other possible combinations/hybridisations of the previous definitions, that form a *taxonomic dictionary* that is varied but also (paradoxically) homogeneous. Summing up a very broad debate (which is also prolific in terms of published results), but also intentionally simplifying common elements, it is not difficult to maintain that all the interpretive standards that we have just mentioned can be attributed to the disappearance of two interpretive categories/metaphors of (urban) space: the centre and the boundary. At the same time, and paradoxically, the pervasive nature of urban space, and the role it plays for a large range of institutions, organisations, subjects and groups, dematerialises and confuses outlines and boundaries, which have become "unimaginably confused" (Brenner 2016). The proliferation of borders, their prismatic break-down and re-composition, is *the other side of globalisation*, both on a micro level of *daily* urban spaces, and on a macro level of intercontinental global flows (Mezzadra 2004). They are conventional and geographic, abstract and real boundaries, that define (and limit) social spaces and phenomena: boundaries that change frequently in space and time, each time including and excluding individuals and places, according to choice or need.

The research carried out, of which this book publishes the results, has addressed this topic, identifying a taxonomy to describe the rules of this new, particular *city* but using the living dimension as a device for the epistemological breaking down of traditional socio-spatial analysis models. This task is not simple, however, as the matter of boundaries/limits of the modern dimension/urban matter includes considerable contradictions and aporias. These changes, and consequent set ups, require an effort to rethink, both regarding the interpretive aspects but mostly the areas of public intervention for the definition of policies and actions that are not sporadic, ineffective or merely repressive. Taking on the challenge of liveability as a prerogative for building inclusive urban territories provides innovation in the field of urban-planning practices, reformulating the idea of well-being/welfare and also of citizenship (Paba and Perrone 2004; Lo Piccolo 2010) and common good (Paba 2003): the analyses carried out in the book show an extremely complex picture of new inhabitants, new demands/ways of living, new citizenship, against a number of *ways* of living, including an informal one, where the policies and disciplinary tools cannot provide answers and often do not result in awareness.

There is a high rate of literature regarding the subject of how difficult it is to reconcile citizens' rights and the plurality of inhabitants (Young 1990; Smith 1996; Sandercock 1998, 2000; Mitchell 2003). The many ways of living require more analysis and responses that can address the matter of inclusion as well as well-being and safety, not opposed to—as indeed happens (Sandercock 2000; England and Simon 2010; Kern 2008)—or with the progressive reduction of public space (Mitchell 2003; Glasze et al. 2006). These phenomena do not only concern global cities: new geographies tied to the effects of urban regionalisation can be seen throughout Southern Europe, including the South of Italy.

From a disciplinary point of view, the relations between the new forms of western cities and the types of living have been investigated for over twenty years (Sorkin 1992; Amendola 1997; Bauman 2000; Dehaene and De Cauter 2008) producing a significant portion of the theoretical reasoning about cities, using the filter of a global north, while the cities in the south continue to be considered as *marginal, less global* and *delayed* compared to the phenomena, practices and processes. Undoubtedly, the urban contexts of the south, and of Southern Europe in particular, are particularly *disorganised* (Malheiros 2002) and the renewal of urban governance practices has been slowed down by bureaucratic networks and institutional inertia (Seixas and Albet 2010). Sicily is thought to be at the limit of this *marginality* and, for many reasons, it is (Cannarozzo 2000; Lo Piccolo 2009; Rossi Doria 2003; Rossi Doria et al. 2005). However, as suggested by Brand and Gaffikin (2007, 284) "hard cases can illuminate the challenges and contradictions involved in a proposition, without laying claim to being typical". From this point of view, the Sicilian urban phenomenon represents an extreme case and the consequent reflections can be helpful in order to individuate and discuss some broader phenomena that are much more evident—and consequently analysable—in *extreme* conditions.

The excess urbanisation of the Sicilian territory is a consolidated phenomenon over a long period of time and takes on different characteristics depending on the contexts, but with similar results in the consumption of soil, whether it is for residential expansion, the increase in holiday homes or tertiary structures for commerce or leisure. Consolidate analyses, still substantially applicable today, have highlighted these distinctive traits, acknowledging "a growth that is not necessarily excessive, but that is abnormal in a city of the south, with a clear obsolescence of any interpretive model of the hierarchy method of the cities" (Becchi Collidà 1978, 43). In Sicily, this phenomenon is predominantly evident in the south, due to reasons that transcend the crumbling of agriculture and the crisis of the territorial production systems. General reflections are needed on development models and on contradictions between the latter and local realities distinguished by marginal, backward economic systems, even in the dynamic and apparently innovative aspects, so the new geographies that have emerged can be adequately represented.

Starting with an analysis and interpretation of urban and territorial transformations, the different outcomes of urban regionalisation processes in a *marginal* context like Sicily have been identified and described. Sicily does not have a history of a true metropolitan stage, both in its physical set-up—including demographic, social and functional—and also regarding development models and economic processes in the second half of the twentieth century (see chapter by Picone et al.).

Due to a limit of time and resources, the investigation did not include the entire island territory, which is extremely complex and heterogeneous in its dynamics and processes. We therefore chose to investigate two territorial contexts that differ greatly from each other, both in terms of spatial location and in characteristics and phenomena: the context around Palermo and the context of South-Eastern Sicily. The *comparison*, and therefore the proposal to work on the two territorial contexts side by side, comes from a working hypothesis that critically addresses the local-regional variation of the metropolitan *model*, in order to monitor the various trajectories of change in relation to the more specific issue of the living demands. Without resorting to the analyses described in later chapters of this book, it is necessary to anticipate how urbanisation processes of the two contexts show very different characteristics, the result of socio-economic phenomena and urban-administrative choices that are the opposite of each other. Reinterpreting this phenomenon in the two contexts, the dynamics of the South-Eastern Sicily context are evident, as it is historically multi-centric and spatially specialised compared to the mono-nuclear character of the Palermo area. The empirical evidence of the dynamic nature of South-Eastern Sicily has been the very element—as we will see in later chapters—on which on one side the critiques of several urban theories have been built, and on the other the pivot element for identifying a specific nature of the dynamics of regionalisation of the marginal contexts that Sicily is a representative sample of.

However, it was speculated that only an accurate analysis free of any territorial stereotypes—that are unfortunately common when the Sicilian territory is being analysed—could confirm or deny the directions of urban and territorial development, with a view to post-metropolitan analysis. Complex dynamics have thus emerged, hidden behind the main socio-economic changes described by statistical data. To adequately understand these dynamics, the research identified less evident social mechanisms, responsible for the most appreciable urban changes, investigating the areas of *what* and *where*, or rather the complex link between the context's social and economic conditions and choices/strategies for use of the territory by the administrators and inhabitants. Starting with the high state of complexity that characterises the two selected study areas, the research focused its attention on the topics of living, diversity, social inclusion and new economic-spatial set-ups, believing that these areas of investigation are useful to provide targeted (and possibly strategic) indications for the required public policies to be enacted not just on the islands of the Italian Mezzogiorno, but also in other Southern European contexts.

The critical reflections on the links between theoretical formulations and phenomena below were the theoretical framework of the research and the basis for the development of the empirical investigation of the urban contexts being studied.

The research has, in some cases, highlighted the unique and specific nature of the post-metropolitan phenomenon in Sicily compared to topics and conceptual processing going on overseas.

In tracking the bibliographical path, this introductory chapter addresses the huge bibliography on these issues without any demand for systematic approach, but pursuing the aim of tracing the geography of the on-going urban regionalisation processes transition in Sicily.

2   The Dimensions of *Space* in the Urban Regionalisation Processes

The interpretations of the growth in urbanisation outside the traditional urban body show how this phenomenon is no longer a marginal, sporadic episode, but an element that is a part of the transformation dynamics of modern urban realities. This unlimited growth is radically changing the urban structures (from a physical and morphological point of view), together with the urban image (from a political and social point of view).

Expressions that have become idiomatic, such as "megalopolis" (Gottmann 1961), "city region" (De Carlo 1962), "linear-city" (Soria y Mata 1968), "ville éparpilleé" (Bauer and Roux 1976), "widespread city" (Indovina et al. 1990), "ecopolis" (Magnaghi 1980), "edge city" (Garreau 1991), "global city-region" (Scott 2001), "suburbia" (Hayden 2003), "mega-city region" (Hall and Pain 2006), "planetary urbanisation" (Brenner 2014) or, with reference to the urban diffusion phenomenon (Ardigò 1967), "counter-urbanisation" (Berry 1976), "disurbanisation" (van den Berg et al. 1982), "reurbanisation" (Bauer 1993), have been useful to tracking the elements of spatial declination in the urban regionalisation processes. At the same time, the analysis of results of the intense period of European studies on the urban phenomenon and the morpho-typological characteristics (Borachia et al. 1988; Astengo and Nucci 1990; Secchi 1993; Clementi et al. 1996; Font 2004; Couch et al. 2007) has allowed us to recognise the process nature of the urban regionalisation nature, that, although with various spatial and formal characteristics, combines and invests contemporary territories with effects on environmental systems (see chapter by Schilleci et al.). In light of this, it is however necessary to underline that if the densely-constructed city still maintains several characteristics of collective construction, of that *civitas* that has informed European urban culture, the huge post-metropolitan territories are increasingly becoming a kind of "counterspace" (Maciocco 2006) where the elements of modernity, that are essentially anti-urban and self-referential, are taking shape.

In the urban regionalisation areas, an internal security is opposed by an external insecurity, feeding a defence rhetoric that creates types of *post-socialisation* between homogeneous groups and sets off spatial mechanisms and devices of exclusion. The new interpretation of the ratio between public space and private space has therefore been an opportunity to identify new theoretical categories of urban space that go beyond the administrative limits of the traditional city (on aspects related to the redefinition of the administrative limits of metropolitan areas in Sicily see chapter by Lotta).

As highlighted by Kohn (2004), the term public space is a cluster concept as it has multiple and sometimes contradictory definitions that, due to the effect of contamination and hybridisation between public realms and private realms, is becoming increasingly difficult to classify. In addition to this problematic nature, there is a *narrative of loss* over public space that is linked to an ideal, abstract concept of the public sphere as an inclusive space where social interaction, political action and cultural exchange unfurl.

To further complicate the public space study, there is also the problem of the dematerialisation of the public sphere, not just from the scientific point of view in terms of sub-theorisation of the public sphere's spatial structure (Low 2004), but also as a progressive decline of the public dimension linked to physical spaces found in society and in modern cities (Sennett 1977; Putnam 1993).

A substantial change in that intermediate realm between State and the private realm has emerged from the above debates, which is the public sphere, in fact (Sebastiani 2007), an element that confers the status of *place* on the space and therefore of public space. And while for Habermas (1989) the public sphere is not a physical place, but rather a non-institutional space where civil society, or rather the *public dimension of private subjects* (Habermas 1962) carries out critical functions in a colloquial form and in this sense carries out political functions (Goheen 1998; Paddison and Sharp 2007; Sebastiani 2007) in post-metropolitan territories, the weakening of the public sphere and the absence of institutional public spaces feed and strengthen a self-referential private sphere, incapable of developing through comparison, or conflict, with *the other*. The relational dimension of public space and collective and common spaces are both missing in the post-metropolitan space.

The private sphere has obviated the absence of this type of public spaces, creating new mono-functional, specialised spaces that are controlled and managed with the typical mechanisms of private ownership (i.e. limiting accessibility over time and usability to some categories of subjects). Shopping malls and hyper-specialised centres in post-metropolises have set off a privatisation process of public space (Mitchell 1995; Smith 1996; Low 2000) based on policies and projects founded not on the principle of public good but on the one of maximum profit.

The privatisation process that affects the post-metropolitan territories does not only take place with the proposition of pseudo-public spaces as an alternative to traditional public spaces, but also in increasingly segregated and closed forms of living. And while at the start of the sub-urbanisation process living forms were found in the isolated villa with garden, in recent decades the emerging trend is linked to the growing spread of the so-called gated communities.

Faced with the changes in lifestyles, with the gap between the city of the rich and the city of the poor (Secchi 2013), with the growing fear of diversity, new types of homogeneous and selective communities were formed in the large post-metropolis, which exclude themselves in their "Privatopia" (McKenzie 1994).

And while in America, the *privatopia option* is a choice connected to matters of security and social homogeneity, in Europe, and in particular in the southern area, self-sufficient residential enclaves obviate the indifference and *policy of non-intervention* that is practised towards urban regionalisation processes.

The gated communities are one of the products of that *separateness urban planning* (Sernini 1997) that finds its archetype in *closed* residential urban area, via natural barriers or artificial obstacles.

In spite of the fact that there is no single definition of these forms of residence, several authors (Blakely and Snyder 1997; Low 2003; Le Goix 2005; Vesselinov et al. 2007) agree in defining them as residential complexes closed off by physical barriers and limited access, where there are various forms of surveillance and security systems (closed-circuit television cameras, guards, etc.) and that usually offer public services and amenities in a private format). According to these authors, this definition is sufficient to explain the idea of living expressed by the emerging residential forms.

The residential model subtended to them produces many questions about the "right to the city" (Lefebvre 1968) in the post-metropolis, feeding forms of private governance and exclusion mechanisms and marginalisation among various groups, that are not only physical barriers but also an emphasis of spatial inequalities in terms of access to spaces and services between the *haves* and *have nots* (Marcuse 2009; Lo Piccolo 2012; Porcu 2013).

What is most concerning is that at the start of the 90s this model—in its various forms—considered to be marginal and sporadic and characteristics of economically advanced territorial contexts, has becoming increasingly successful, making itself the desirable and hoped for dimension, even in our contexts that are historically considered to be at the margins of said processes. As highlighted in chapter by Giampino, these dynamics are changing urban space in Sicily too, via different spatial traditions, but also via new uses of the territory.

Lastly, a further consequence of the urban regionalisation processes is connected to the central role assumed by the urban image in the definition of what a post-metropolis is today. The famous "Precession of Simulacra" by Jean Baudrillard (1983) is a founding text for a broad debate on the relationship between urban studies and the ones that we now call visual studies (Bignante 2011), but still a study from over thirty years ago. It is unnecessary to point out how, in the meanwhile, an increasing number of reflections have been published about the city image (starting with the forerunner works of Lynch 1960; Venturi et al. 1972) and its increasing dramatisation (Minca 2005), on the relationship between capitalism and city image (Short and Kim 1999; Rossi and Vanolo 2010), on the virtual (Lévy 1995), on the bond between city and cinema (Sandercock and Attili 2010; Shiel and Fitzmaurice 2011), on "urban gamification" (Olthof and Eliëns 2017; Vanolo 2018) and so on. Soja (2000), for example, after discussing the contributions of the "Precession of Simulacra" by Jean Baudrillard (1983) and the "Psychasthenia" by

Celeste Olalquiaga (1992) (much less known in Italy than the work by the French philosopher), introduces a term into his speech about "Simcities" that was extremely fashionable between the twentieth and twenty-first centuries, but that has now lost a large part of its popularity, apart from some recent revivals (also in the cinema): the concept of "cyberspace".

"Cyberspace" is a literary invention linked to the novel "Neuromancer" by William Gibson (1984), but reflects the cultural climate of the 1980s, which foretold of a drastic change in the very idea of a city (Picone 2006). The best-known exponent of this climate in the cinema is the film "Blade Runner" by Ridley Scott (1982). Inspired by a *cyberpunk* novel by Philip K. Dick, "Blade Runner" may seem like a banal science fiction film, but if authors like Harvey (1989), Davis (1992), Boyer (1996) and Soja highlight the link between this film and the new city model that Los Angeles is the standard bearer of, it means that there are elements worthy of consideration without a doubt. These considerations made us reflect, as we can see in later chapters (in particular see chapter by Todaro et al.), on the way in which Palermo and South-Eastern Sicily have modified and continue to modify their urban image. Of course, these changes in image also pass through the power of television fiction series (Lo Piccolo et al. 2015), but not just that.

3 Globalisation and Fragmentation of Urban Space

The neoliberal *restructuring* of the post-modern city (Brenner and Theodore 2002), is inextricably connected to multiform and multilevel urban transformation phenomena connected with globalisation processes.

This unmistakeable conceptualisation hides a rather complex observation, that may be controversial and sometimes contradictory, that increases around globalisation, *an apparently all-inclusive metaphor* (Soja 2000) and on the effects this produces in terms of social and spatial fragmentation of urban space.

This complexity helps us to grasp the joint existence—that is apparently impossible to understand—of the processes and phenomena of spatial standardisation/differentiation and social equality/inequality; but at the same time to become more aware of the significant consequences in terms of socio-spatial stratification, fragmentation and segregation, and the matters of power and the consequent types of social inequality that come from globalisation processes.

The topic of urban space globalisation is addressed at length in scientific literature, and in urban studies (Short and Kim 1999; Brenner 2004), and also in political economics (Savitch and Kantor 2002), urban sociology (Bagnasco and Le Galès 2000; Le Galès 2002) and in socio-economic contexts (Friedmann 1986; Jameson 1991; Sassen 1991, 1994).

Soja (2000) includes his arguments in the mainstream narrative on the Keynesian-Fordist model founded historically on the connections between the accumulation of capital/work, the Keynesian welfare state and new urbanisation phenomena. Its decline, marked by the end of the *Golden Age* (Fordism) and the 1970s international crisis, generated a new world economic system (Badie 1995) defined by a multilevel governance of economic and political institutions (Held 2004) that act on a global and local scale simultaneously (Scott 2001). Cities became protagonists of this new system (d'Albergo and Lefèvre 2007), while a generally more blurred, marginal profile emerged for national states (Paddison 2001; Beck 2005).

In relation to this new set-up, for some authors like Hall (1996), Friedmann and Wolff (1982) and Friedmann (1986) with the "world cities", or Sassen (1991) with the "global cities", cities are still privileged interlocutors in global economic relations, while for others like Gottmann (1961) with the "megalopolis", Scott (1996, 2001) with the "global city-regions" or Hall and Pain (2006) with the "global megacity-regions", the regions are the new engines of the world economy ("world of regions", Scott 2001): on a regional scale, cities are grouped into spatially polarised coalitions, in an attempt to address globalisation risks and opportunities more effectively.

Brenner and Schmid's (2014, 2015) most recent analysis proposes the overcoming of the previous paradigms, according to a new interpretation that recognises a theoretical category and not an empirical subject in *urban* areas. Such theoretical conceptualisation denies almost any spatial limit to cities, referring to "planetary urbanisation" (Brenner 2004; Brenner and Schmid 2011). This is the result of concentrated, widespread and differential urbanisation processes, where it is no longer possible to draw the boundary or the difference between urban and rural. According to Brenner and Theodore (2002), in fact, the modern urban issue conforms to the unequal extension of the "spatial and institutional creative destruction process" on global scale (according to Lefebvre's theory of space, 1992) that involved urban bodies rather than the formation of a worldwide network of global cities or a worldwide megalopolis.

Going beyond the reasoning on scales and degrees of action, however, the main fields of application analysed in the city reorganisation process in the globalisation era, i.e. economic (Friedmann and Wolff 1982; Friedmann 1986; Harvey 1989; Jameson 1991; Sassen 1991) and socio-spatial (Soja 2000), did not sufficiently investigate the socio-cultural aspects with which both the macro- and the micro-economic transformations and the spatial changes are directly related to. What happens inside cities and global regions? What are the social and cultural components affected by these new phenomena? Saskia Sassen (1991) spoke of these aspects, addressing her research to the matters of power and social inequality coming from globalisation processes. Also, Engin Isin (1991), by comparing "cosmopolitan social theory" and planning, investigated the modern battle for citizenship rights and rights to the city, underlining how the cosmopolis is an arena where new forms of citizenship are created.

Towards the end of the 1990s, in the transition from metropolis to cosmopolis, Leonie Sandercock (1998) proposes an upside down interpretation, wherein the current urban condition (cosmopolis) becomes the scenario where some of the main emerging socio-cultural forces (civil society, migrations, feminism and the demands of other historically oppressed groups) produce transformations, anxieties, conflicts, according to a fear of difference that creates social fragmentation in urban space.

The reflections on a fractal city by Soja (2000) range from the awareness of the intensification of these socio-economic and spatial inequalities in the urbanisation processes that can be traced to the new post-Fordist geographies. The fluidity and complexity of its restructured social mosaic is the founding epistemological assumption. This condition is accompanied by the inadequacy of the traditional dualistic models (rich and poor, white and black, men and women, etc.) for interpreting the typical polarisation of a Fordist society (Mollenkopf and Castells 1991) and the need for more suitable analytical tools.

One of the most significant facts, therefore, is, in spatial terms, the growth and diversification of the polarisation phenomena of poverty and wealth, and the consequent increase in the urban distribution of socio-economic differences/inequalities (Fincher and Jacobs 1998).

Compared to such references, the contributions that concentrate on the effects of new information technology on socio-spatial fragmentation in post-Fordist cities (Castells and Hall 1994; Castells 1996; Ellin 1996; Graham and Marvin 1996, 2001; Scott and Soja 1996; Curry 1998; Friedman 1999) acknowledge a decisive role for the new technologies in renovating urban spaces; and also in the production of the parallel phenomena of fragmentation and recombination of uses and functions, inside and between cities and city systems (Mitchell 1996). As underlined by Manuel Castells (1996), these phenomena support the consolidation of an urbanised network society, which is increasingly globally integrated but fragmented at the same time, creating new highly polarised urban landscapes where the Information Technology, Electronics and Communications (ITEC) networks selectively connect the pre-chosen users and the places, excluding all the rest.

These studies are of course accompanied by political-economic contributions (Sassen 1991; Brenner and Theodore 2002; Peck and Tickell 2002; Harvey 2005) that connect these processes with the neo-liberalisation of the world economy at the end of the 1970s in a more orthodox manner; or the contributions of social research (Kofman 1995; Allen and Turner 1997) that specifically analyse the new social-hybridisation models in post-metropolis space, with special reference to different ethnic groups and immigrants.

Following these interpretations, the first results of the increase in socio-spatial fragmentation in globalised cities can be traced to different conceptualisations, such as "quartered city" (Marcuse 1989), "archipelago city" (Davis 1990), "divided cities" (Fainstein et al. 1992), "partitioned city" (Marcuse 1995; Marcuse and van Kempen 2000), "metropolarities" (Soja 2000). Also, these refer to heterogeneous spatial visions, they maintain the selective and segregating character of *fragments* of space for increasingly exclusive uses, and therefore tend to create widespread phenomena of social exclusion (Graham and Marvin 1996, 2001).

Spatial polarisation, differentiation of socio-economic inequalities and new exclusion geographies are thus the components that catalyse the fight for respecting plurality (Arendt 1958) and for the full recognition of the "right to difference" (Young 1990; Soja 2000); these lead to widespread rebellious practices of citizenship (Friedmann 1987; Holston 1995), creating new challenges for planning tools (Sandercock 2000). These movements recognise space as the subject of social conflict, and at the same time as the place of its representation (Mitchell 2003; Harvey 2012). The social production of new spatiality thus becomes an active part in the reproduction of inequalities and forms of social injustice, but at the same time a tool for objecting to them.

The repercussions of these reflections on the Sicilian regional context on which this study is focused are controversial. On the one hand, in its clear distance due to the nature of the phenomena, the specific context and scale of analysis, Sicily shows new and somewhat surprising representations. In this sense, overcoming the rhetoric of *post-Fordism without Fordism?* also extended to the post-metropolitan transition without a complete metropolitan stage, has allowed us to look at the Sicilian territory with a rather disenchanted view, allowing us to grasp certain specifics that cast doubt on the assertive nature of mainstream visions.

On the other hand, polarisation of socio-economic inequalities and socio-spatial fragmentation find representations in Sicily—traditionally considered to be marginal contexts—that are very close to those of globalised contexts.

More specifically, reflections on the concentration/dispersion of the foreign population compared to a territory's specificity (see chapter by Todaro and Lo Piccolo), together with cases of spatial segregation of the same immigrants that can be referred in particular to urban contexts (see chapter by Busetta et al.), the right to housing (see chapter by Giampino et al.) and the new forms of living in the globalisation era (see chapter by Abbate), or the threshold income analysis, used for mapping the capacity/impossibility of buying a house (see chapter by Bonafede and Napoli), are just some of the emerging issues that place Sicilian territories not too distantly from the geographical areas that literature on this topic considers to be reference contexts.

4   Conclusions

In light of the reflections that have emerged, the chapters below weave a non-conventional narration, which is at times atypical and divergent when compared to the reference theoretical models mentioned several times, contributing to question the reasons. This book tries to answer basic issues and the questions explained above, for which the interpretations provided in reference literature are not totally convincing. In fact, these references leave no room for different points of view, and refer to a widespread dominant perspective that we don't believe restores the post-modern complexity of urban space to a sufficient degree (Giampino et al. 2014).

In reference to the research situation, these considerations were further substantiated by the choice of working according to a comparative approach, aimed at critically reflecting on urban space (Balducci et al. 2017), on a dual field of observation: an external one, that compared *dominant* theoretical models and Sicily, with its condition of *marginality*; and an internal one, between the Metropolitan Area of Palermo and South-Eastern Sicily, with its different specificities, which can however be traced to more general issues and models.

In the former case, the research responds to the need to not look at territories *from north to south*, but turning the traditional observation axis upside down, in order to be able to fully understand the possibly different dimension of the marginal areas, that are not generally affected by the clear phenomena of metropolitanisation (Giampino et al. 2014). This then contributes to relativising the dominant nature of the reference models, adding new, heterogeneous *socio-spatial geographies* that are the most important outcome of the research.

In the latter case, the comparison of the Palermo context, which is traditionally considered to be a metropolitan area starting with its clear monocentrism, and South-Eastern Sicily, which is more dynamic and innovative in a socio-economic sense, in spite of its territorial profile that is spatially multicentric and more fragmented, casts doubt on the traditional belief that modern conditions only travel through a metropolitan phase.

In both cases, some of the metaphors used as a reference—for example the metaphor of the post-metropolis—have not been taken on as new urban forms, or as models, but as interpretive tools for understanding the current changes in urban regions (Soja 2015). In this sense, it would not be useful to describe the new forms and morphotypes of urban space present today but rather restore the characteristics of change and differentiation, the result of ongoing urbanisation processes, compared to codified models and forms.

Therefore, the book starts with the referred to theoretical reasoning and refers to an apparently atypical territory when compared to some of the interpretations offered. By proposing a critical reflection on urban regionalisation processes that affect the two areas of research in different ways, the book is split into three parts.

The first part contains a reflection on socio-spatial phenomena that the urban regionalisation processes bring with them. In this sense, the quantitative aspects of the socio-economic aspects of the analysed contexts are compared with the spatial phenomena of urban settlement transformation. The latter are interpreted using forms of settlement pressure carried out on environmental systems. This part ends with a circumstantial analysis of the types of spatial segregation of the foreign population in Palermo, the largest city on the island.

The second part addresses the central topic of informal practices that are an increasing part of urban transformation/evolution processes with methods and approaches that thoroughly reject the traditional models of intervention, placing planning before new ethical challenges. The main topic of claiming rights is explained on a spatial front, with aspects that can be traced to the new issue of housing and the right to a home, and on a social one, in reference to the condition of the *invisibility* of immigrants in the quality agricultural productions in rural areas. In the case of South-Eastern Sicily, the consequent spatial and social effects produced by the sometimes-controversial relationship between media image and territorial reality are discussed, with regard to the *optimisation policies* of territorial assets.

Lastly, the third part addresses matters that can be directly traced to the sphere of institutional policies, proposing a reflection on how they need to evolve in the direction of a more suitable refinement of knowledge-gathering tools, and also in redefining the models and relative governance tools for the transformations taking place. This reflection begins with reasoning that starts from the evolution of the forms of (post-)metropolitan dimension government and then asks questions about the more suitable tools for managing the new urban forms. It then compares the anomaly of current urban regionalisation processes with the lack of effective governance policies and ends with the definition of income threshold analysis models, aimed at defining the most suitable fact-finding tools for mapping the capacity/impossibility of buying a house.

In relation to these fields of observation, therefore, we have attempted to explain the convergences and divergences, the agreements and disagreements in order to contribute to the wealth of critical reflection on urban space (Brenner 2016) with a view to a different and not standardised interpretation (Roy 2009).

Research results portray a story that is balanced between the already-mentioned *concrete abstraction* (Brenner 2016) of the new urban dimension, and the *marginality* and boundary of a unique, new territorial reality, such as Sicily, that can question the theoretical reference models.

References

Allen JP, Turner E (1997) The ethnic quilt: population diversity in Southern California. Northridge, Orem UT

Amendola G (1997) La città postmoderna. Magie e paure della metropoli contemporanea. Laterza, Rome-Bari

Arendt H (1958) The human condition. University of Chicago Press, Chicago

Ardigò A (1967) La diffusione urbana. Le aree metropolitane e i problemi del loro sviluppo. Editrice AVE, Rome

Astengo G, Nucci C (1990) It.Urb. 80: rapporto sullo stato dell'urbanizzazione in Italia. Urbanistica Quaderni 8

Badie B (1995) La fin des territoires. Essai sur le désordre international et sur l'utilité sociale du respect. Fayard, Paris

Bagnasco A, Le Galès P (2000) Cities in contemporary Europe. Cambridge University Press, Cambridge

Balducci A, Fedeli V (a cura di) (2007) I territori della città in trasformazione. Tattiche e percorsi di ricerca. FrancoAngeli, Milan

Balducci A, Fedeli V, Curci F (eds) (2017) Post-metropolitan territories: looking for a new urbanity. Routledge, London

Baudrillard J (1983) The precession of simulacra. In: Baudrillard J (ed) Simulations. Semiotext(e), New York, pp 1–79

Bauer I (1993) Les suburbia, sommes-nous concernés? Urbanisme 1:67–88

Bauer G, Roux JM (1976) La rurbanisation ou ville éparpillée. Edition du Seuil, Paris

Bauman Z (2000) Liquid modernity. Polity, Cambridge

Becchi Collidà A (1978) La città meridionale. In: Indovina F (ed) Mezzogiorno e crisi. Situazione economica, struttura urbana, conflitti e forze politiche. FrancoAngeli, Milan, pp 41–96

Beck U (2005) Lo sguardo cosmopolita. Carocci, Rome

Berry BJL (ed) (1976) Urbanization and counter-urbanization. Sage, London

Bignante E (2011) Geografia e ricerca visuale. Strumenti e metodi. Laterza, Rome-Bari

Blakely E, Snyder MG (1997) Fortress America. Gated communities in the United States. Brooking Institution Press, Washington

Borachia V, Moretti A, Paolillo PL, Tosi A (a cura di) (1988) Il parametro suolo. Dalla misura del controllo alle politiche di utilizzo. Grafo, Brescia

Boyer MC (1996) CyberCities: visual perception in the age of electronic communication. Princeton Architectural Press, Princeton

Brand R, Gaffikin F (2007) Collaborative planning in an uncollaborative world. Plann Theory 6(3):282–313

Brenner N (2004) New state spaces: urban governance and the rescaling of statehood. Oxford University Press, Oxford

Brenner N (2014) Introduction: urban theory without an outside. In: Brenner N (ed) Implosions/Explosions: towards a study of planetary urbanization. Jovis, Berlin, pp 14–33

Brenner N (2016) Stato, spazio, urbanizzazione. Guerini e Associati, Milan

Brenner N, Schmid C (2011) Planetary urbanization. In: Gandy M (ed) Urban constellations. Jovis, Berlin, pp 10–13

Brenner N, Schmid C (2014) The "urban age" in question. Int J Urban Reg Res 38(3):731–755

Brenner N, Schmid C (2015) Towards a new epistemology of the urban? City 19(2–3):151–182

Brenner N, Theodore N (eds) (2002) Spaces of neoliberalism: urban restructuring in North America and Western Europe. Blackwell, Malden

Cannarozzo T (2000) Palermo: le trasformazioni di mezzo secolo. Archivio di Studi Urbani e Regionali 67:101–139

Castells M (1996) The rise of the network society. The information age, economy, society and culture I. Blackwell, Oxford

Castells M, Hall P (1994) Technopoles of the world: the making of twenty-first century industrial complexes. Routledge, London

Clementi A, Dematteis G, Palermo PC (eds) (1996) Le forme del territorio italiano, vol I, Temi e immagini del mutamento. Laterza, Bari

Couch C, Leontidou L, Petschel-Held G (ed) (2007) Urban sprawl in Europe: landascapes, land-use change & policy. Blackwell, Oxford

Curry M (1998) Digital places: living with geographic information technologies. Routledge, London

d'Albergo E, Lefèvre C (2007) Come e perché le città vanno all'estero. In: d'Albergo E, Lefèvre C (eds) Le strategie internazionali delle città. Dieci metropoli a confronto. il Mulino, Bologna, pp 9–39

Davis M (1990) City of quartz: excavating the future in Los Angeles. Verso, London

Davis M (1992) Beyond blade runner: urban control, the ecology of fear. Open Magazine Pamphlet Series, New Jersey

De Carlo G (1962) La nuova dimensione della città. La città regione. ILSES, Stresa

Dear M, Flusty S (1998) Postmodern urbanism. Ann Assoc Am Geogr 88(1):50–72

Dehaene M, De Cauter L (eds) (2008) Heterotopia and the city: public space in a postcivil society. Routledge, Abingdon

Ellin N (1996) Postmodern urbanism. Blackwell, Oxford

England M, Simon S (2010) Scary cities: urban geographies of fear, difference and belonging. Soc Cult Geogr 11(3):201–207

Fainstein S, Gordon I, Harloe M (eds) (1992) Divided cities. Blackwell, Oxford

Fincher R, Jacobs JM (1998) Cities of difference. Guilford, New York

Font A (ed) (2004) L'explosió de la ciutat-Morfologies, mirades y mocions. COAC publicacions, Barcelona

Friedman K (1999) Restructuring the city: thoughts on urban patterns in the information society. The Swedish Institute for Future Studies, Stockholm

Friedmann J (1986) The world cities hypothesis. Dev Change 17:69–83

Friedmann J (1987) Planning in the public domain: from knowledge to action. Princeton University Press, Princeton, NJ

Friedmann J, Wolff G (1982) World city formation: an agenda for research and action. Int J Urban and Reg Res XV(1):269–283

Garreau J (1991) Edge city: life of the new frontier. Anchor Books, New York

Giampino A, Picone M, Todaro V (2014) Postmetropoli in contesti al margine. In: Atti della XVII Conferenza Nazionale SIU, L'urbanistica italiana nel mondo, 15–16 May 2014, Milan. Planum. J Urbanism 2(29):1308–1316

Gibson W (1984) Neuromancer. The Ace Publishing Group, New York

Glasze G, Webster C, Frantz K (eds) (2006) Private cities: global and local perspectives. Routledge, London

Goheen PG (1998) Public space and the geography of the modern city. Prog Hum Geogr 22(4):479–496

Gottmann J (1961) Megalopolis: the urbanized northeastern seaboard of the United States. The Twentieth Century Fund, New York

Graham S, Marvin S (1996) Telecommunications and the city: electronic spaces, urban places. Routledge, London

Graham S, Marvin S (2001) Splintering urbanism: networked infrastructures, technological mobilities, and the urban condition. Routledge, London

Habermas J (1962) Strukturwandel der Öffentlichkeit. Untershungen zu einer Kategorie der bürgrlichen Gesellschaft. Suhrkamp Verlag, Frankfurt am Main

Habermas J (1989) The structural transformation of the public sphere: an inquiry into a category of Bourgeois Society. Polity Press, Cambridge

Hall P (1996) The world cities. McGraw-Hill, New York

Hall P, Pain K (eds) (2006) The polycentric metropolis: learning from mega-city regions in Europe. Earthscan, London

Harvey D (1989) The condition of postmodernity: an enquiry into the origins of cultural change. Blackwell, Cambridge and Oxford

Harvey D (2005) A brief history of neoliberalism. Oxford University Press, Oxford

Harvey D (2012) The urban roots of financial crises: reclaiming the city for anti-capitalist struggle. Socialist Regist 48(1):1–34

Hayden D (2003) Building suburbia: green fields and urban growth, 1820–2000. Pantheon, New York

Held D (2004) Global covenant: the social democratic alternative to the Washington consensus. Polity Press, Cambridge

Holston J (1995) Spaces of insurgent citizenship. Plann Theory 13:35–52

Indovina F, Matassoni F, Savino M, Sernini M, Torres M, Vettoretto L (1990) La città diffusa. DAEST, Venice

Isin E (1991) Cosmopolis. Rizzoli, Milan

Jameson F (1991) Postmodernism or, The cultural logic of late capitalism. Duke University Press, Durham

Kern K (2008) Heterotopia of the Theme Park street. In: Dehaene M, De Cauter L (eds) Heterotopia and the city: public space in a postcivil society. Routledge, Abingdon, pp 104–115

Kofman E (1995) Citizenship for some but not for others: spaces of citizenship in contemporary Europe. Political Geogr 14(2):121–137

Kohn M (2004) Brave new neighborhoods: the privatization of public space. Routledge, New York and London

Le Galès P (2002) European cities: social conflicts and governance. Oxford University Press, Oxford

Le Goix R (2005) Gated communities: sprawl and social segregation in Southern California. Hous Stud 20(2):323–343

Lefebvre H (1968) Le droit à la ville. Éditions Anthropos, Paris

Lefebvre H (1970) La révolution urbaine. Éditions Gallimard, Paris

Lefebvre H (1992) The production of space. Wiley-Blackwell, Hoboken, NJ

Lévy P (1995) Qu'est-ce que le virtuel? La Découverte, Paris

Lo Piccolo F (2009) Territori agricoli a latitudini meridiane: residui marginali o risorse identitarie? In: Lo Piccolo F (ed) Progettare le identità del territorio. Alinea, Firenze, pp 11–42

Lo Piccolo F (2010) The planning research agenda: plural cities, equity and rights of citizenship. Town Plann Rev 81(6):i–vi

Lo Piccolo F (2012) Nuovi abitanti e diritto alla città: compiti (tecnici) e responsabilità (etiche) della disciplina urbanistica. Planum. J Urbanism 25(2):1–5

Lo Piccolo F, Giampino A, Todaro V (2015) The power of fiction in times of crisis: movie-tourism and heritage planning in Montalbano's places. In: Gospodini A (ed) Proceedings of the International Conference on Changing Cities II: Spatial, Design, Landscape Socio-Economic Dimensions. Porto Heli, Peloponnese, Greece, 22–26 June 2015, Grafima Publ, Thessaloniki, pp 283–292

Low SM (2000) On the plaza: the politics of public space and culture. Texas UP, Austin

Low SM (2003) Behind the gates: life, security and the pursuit of happiness in Fortress America. Routledge, London

Low SM (ed) (2004) Space and democracy: geographical perspectives on citizenship, participation and representation. Sage, London

Lynch K (1960) The image of the city. MIT Press, Cambridge

Maciocco G (2006) Il progetto ambientale nelle aree di bordo. In: Maciocco G, Pittaluga P (eds) Il progetto ambientale nelle aree di bordo. FrancoAngeli, Milan, pp 7–34

Magnaghi A (1980) Ecopolis, per una città di villaggi. Housing, 3

Malheiros J (2002) Ethnicities: residential patterns in Northern European and mediterranean metropolises: implications for policy design. Int J Popul Geog 8(2):89–106

Marcuse P (1989) Dual city: a muddy metaphor for a quartered city. Int J Urban Reg Res 13(4):697–708

Marcuse P (1995) Not chaos but walls: postmodernism and the partitioned city. In: Watson S, Gibson K (eds) Postmodern cities and spaces. Blackwell, Oxford, pp 243–254

Marcuse P (2009) From critical urban theory to the right to the city. City 2(13):185–197

Marcuse P, van Kempen R (2000) Conclusion: a changed spatial order? In: Marcuse P, van Kempen R (eds) Globalizing cities: a new spatial order?. Blackwell, Oxford, pp 249–276

McKenzie E (1994) Privatopia: homeowner associations and the rise of residential private government. Yale University, New Haven

Mezzadra S (2004) Confini, migrazioni, cittadinanza. Scienza & Politica 30:83–92

Minca C (ed) (2005) Lo spettacolo della città. Cedam, Padova

Mitchell D (1995) The end of public space? People's park definition of the public and democracy. Ann Assoc Am Geogr 85:108–133

Mitchell W (1996) City of bits: space, place and the Infobahn. MIT Press, Cambridge, MA

Mitchell D (2003) The right to the city: social justice and the fight for public space. Guilford Press, New York

Mollenkopf J, Castells M (eds) (1991) Dual city: restructuring New York. Russell Sage Foundation, New York

Olalquiaga C (1992) Megalopolis. University of Minnesota Press, Minneapolis

Olthof T, Eliëns A (2017) Doing the right thing: gamification as a means to tuning human behaviour. In: Yamu C, Poplin A, Devisch O, de Roo G (eds) The virtual and the real in planning and urban design. Routledge, London, pp 42–48

Paba G (2003) Movimenti urbani. Pratiche di costruzione sociale della città. FrancoAngeli, Milan

Paba G, Perrone C (eds) (2004) Cittadinanza attiva. Alinea, Florence

Paddison R (2001) Communities in the city. In: Paddison R (ed) Handbook of urban studies. Sage, Thousand Oaks, CA, pp 194–205

Paddison R, Sharp J (2007) Questioning the end of public space: reclaiming control of local banal spaces. Scott Geog J 123(2):87–106

Peck J, Tickell A (2002) Neoliberalizing space. Antipode 34(3):380–404

Picone M (2006) Il ciclo di vita urbano in Sicilia. Rivista Geografica Italiana 113:129–146

Porcu M (2013) Quartieri privati: stato dell'arte e prospettive di ricerca. Cambio 6(III):89–100

Putnam RD (1993) La tradizione civica delle regioni italiane. Mondadori, Milan

Rossi Doria B (2003) La Sicilia: da Regione del Mezzogiorno a periferia dell'Europa 'forte'. In: Lo Piccolo F, Schilleci F (eds) A sud di Brobdingnag. L'identità dei luoghi: per uno sviluppo locale autosostenibile nella Sicilia Occidentale. FrancoAngeli, Milan, pp 11–41

Rossi Doria B, Lo Piccolo F, Schilleci F, Vinci I (2005) Riconoscimento e rappresentazione di fenomeni territoriali inediti in Sicilia. In: Carta M, Leone NG, Ronsivalle D (eds) Atti della IX Conferenza SIU Terre d'Europa e fronti Mediterranei. Zangara, Palermo, vol 1, pp 263–273

Rossi U, Vanolo A (2010) Geografia politica urbana. Laterza, Rome-Bari

Roy A (2009) The 21st-century metropolis: new geographies of theory. Reg Stud 43(6):819–830

Sandercock L (1998) Towards cosmopolis: planning for multicultural cities. Wiley, London

Sandercock L (2000) When strangers become neighbours: managing cities of difference. Plann Theor Pract 1(1):13–30

Sandercock L, Attili G (eds) (2010) Multimedia explorations in urban policy and planning: beyond the flatlands. Springer, Rotterdam

Sassen S (1991) The global city: New York, London, Tokyo. Princeton University Press, Princeton

Sassen S (1994) Cities in a world economy. Pine Forge, Thousand Oaks

Savitch HV, Kantor P (2002) Cities in the international marketplace: the political economy of urban development in North America and Western Europe. Princeton University Press, Princeton, NJ

Scott AJ (1996) Regional motors of the global economy. Futures 28:391–411

Scott AJ (2001) Introduction. In: Scott AJ (ed) Global City-Regions. Trends, Theory, Policy. Oxford University Press, Oxford, pp 1–10

Scott AJ, Soja EW (eds) (1996) The city: Los Angeles and urban theory at the end of the twentieth century. University of California Press, Berkeley and Los Angeles

Sebastiani C (2007) La politica delle città. il Mulino, Bologna

Secchi B (1993) Le trasformazioni dell'habitat urbano. Casabella 600:44–45

Secchi B (2013) La città dei ricchi e la città dei poveri. Laterza, Bari

Seixas J, Albet A (2010) Urban governance in the South of Europe: Cultural identities and global dilemmas. Analise Soc 45(197):771–787

Sennett R (1977) The fall of public man. Knopf, New York

Sernini M (1997) Urbanistica della separatezza/Urbanistica della connessione. Archivio di Studi Urbani e Regionali 59:133–150

Shiel M, Fitzmaurice T (2011) Cinema and the city: film and urban societies in a global context. Wiley, London

Short JR, Kim YH (1999) Globalization and the city. Pearson, Harlow

Smith N (1996) The new urban frontier: gentrification and the revanchist city. Routledge, London

Soja EW (2000) Postmetropolis: critical studies of cities and regions. Blackwell, Malden, MA

Soja EW (2015) Accentuate the regional. Int J Urban Reg Res 39(2):372–381

Soria y Mata A (1968) La città lineare. Il Saggiatore, Milan

Sorkin M (ed) (1992) Variations on a theme park: the new American city and the end of the public space. Hill and Wang, New York

van den Berg L, Drewett R, Klaassen L, Rossi L, Vijverberg C (1982) Urban Europe: a study of growth and decline. Pergamon Press, Oxford

Vanolo A (2018) Cities and the politics of gamification. Cities 74:320–326

Venturi R, Scott Brown D, Izenour S (1972) Learning from Las Vegas. MIT Press, Cambridge

Vesselinov E, Cazessus M, Falk W (2007) Gated communities and spatial inequality. J Urban Aff 29(2):109–127

Young IM (1990) Justice and the politics of difference. Princeton University Press, Princeton

Spatial Phenomena

From the Metropolitan Areas to the Post-metropolitan Dimension

Marco Picone, Francesco Lo Piccolo, and Filippo Schilleci

Abstract This chapter introduces a description of the two main areas that will be discussed throughout the book: the metropolitan region of Palermo and South-Eastern Sicily. The chapter starts with a socio-economic analysis describing the characteristics of demography, employment and income in these two parts of Sicily, by comparing them to the situation of other areas in Italy. A spatial analysis of housing and land use complements the socio-economic outlook. Each section describes the most consolidated trends in these areas, but also discusses the most innovative changes and the challenges Sicily has been facing over the years.

1 Approaching the Post-metropolitan Dimension in Sicily

Throughout this chapter, we are going to introduce the two main areas our research group has analysed in Sicily: the metropolitan region of Palermo and the South-Eastern region. We intend to compare these two Sicilian case studies in order to appraise the similarities and differences between them, but most of all to prove how these two particular cases show some unexpected post-metropolitan traits. This portrait also displays several references to the rest of Italy which can help the reader grasp the uniqueness and peculiarity of the island, as related to the Italian context.

Palermo is the most populated and important city of Sicily, and the fifth most populated city in Italy. Although it is widely known for its marginality (Cannarozzo 2000; Pinzello 2003; Rossi Doria 2003; Rossi Doria et al. 2005; Lo Piccolo 2009; Lo Piccolo et al. 2013) and the presence of criminal organisations (*mafia*), Palermo

M. Picone (✉) · F. Lo Piccolo · F. Schilleci
University of Palermo, Palermo, Italy
e-mail: marco.picone@unipa.it

F. Lo Piccolo
e-mail: francesco.lopiccolo@unipa.it

F. Schilleci
e-mail: filippo.schilleci@unipa.it

F. Lo Piccolo et al. (eds.), *Urban Regionalisation Processes*,
UNIPA Springer Series,
https://doi.org/10.1007/978-3-030-64469-7_2

21

has long played a key role in the Mediterranean basin. One of the traditional clichés connected to Palermo and Sicily is that its people has developed a passive attitude as a consequence of the many foreign dominations of the island (starting from its foundation, Palermo has been ruled by the Carthaginians, the Romans, the Arabs, the Normans, the French, and the Spanish). As for all clichés, this one is partly true, but cannot be considered the only reason that explains the city's complex situation.

Most scholars point out that the city of Palermo has never been a metropolis (Giampino et al. 2014), and therefore there is no way it could be considered a post-metropolis, having almost entirely skipped the metropolitan phase. However, history should not be considered as a straight line connecting the past to the future along the same regular path: there are lots of twisted patterns and recurrences in the case of Sicily. Thus, even Palermo might show a few traits that characterise other post-metropolises. This does not mean that we are claiming Palermo's post-metropolitan nature as an easily acceptable status. However, the recent economic crises are blurring the line between Northern and Southern Italy in many ways, and Palermo seems to be undergoing a new stage in the relationships with its hinterland and the other parts of Italy, as we will show.

Given the marginality of its geographical position, we will present the Palermo region through its *area metropolitana* (metropolitan area), an administrative entity which was introduced in Sicily in 1986 and conceived to include 27 municipalities from Partinico (west) to Termini Imerese (east). These 27 municipalities include Ustica, a very small island northwest of Palermo, which arguably shows peculiar traits and cannot be related to the other municipalities (Piraino 1988; Di Leo and Esposito 1991; Grasso 1994).

The *area metropolitana*, although expected to be operative since 1995, has actually never played an active role in this region (Di Leo 1997; Schilleci 2005, 2008a, b). The *Regione Siciliana* (the administrative regional authority) has recently worked on a reform that should lead to the full establishment of the *Città Metropolitana* (Metropolitan City) of Palermo (Lotta 2015).

We have chosen to consider the metropolitan area in order to take into account the relationships linking Palermo to its hinterland, but we have to stress that this condition is quite different from other cases in Italy, as Western Sicily is largely surrounded by sea and has only minor relationships with the rest of Italy. The Palermo region embraces the Metropolitan Area of Palermo (Fig. 1).

The other area we are discussing in this chapter, as previously mentioned, is South-Eastern Sicily (SES). This area is apparently the opposite of any likely post-metropolitan case study. It holds no major city (with the partial exception of Syracuse and its 118,000 inhabitants) and appears to be an island within an island, being one of the most marginal areas of all Sicily (Nobile 1990; Schilirò 2012). Media representations have contributed to strengthen this imagery, by portraying South-Eastern Sicily as a land that is lost in the past, in the echoes of the Baroque era and traditional agriculture (Cannarozzo 2010; Abbate 2011; Azzolina et al. 2012; Lo Piccolo et al. 2015), with people hardly speaking Italian at all. However, this is far from the truth. South-Eastern Sicily is probably one of the most dynamic areas of Southern Italy (Asso and Trigilia 2010) and shows some post-metropolitan traits that

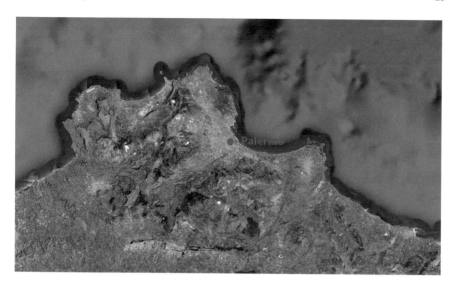

Fig. 1 The Metropolitan Area of Palermo (https://www.google.it/maps/)

are definitely worth discussing. One might consider SES as a *counter-case* to the true post-metropolitan Italian cases like Milan and Turin; however, this counter-case has much to say on how the post-metropolitan nature is not necessarily tied to megacities or to a world-leading economy. Within this chapter, we will try to deconstruct the clichés related to South-Eastern Sicily and describe the most innovative trends one can find in this remote corner of Italy.

South-Eastern Sicily, according to our analysis, embraces nineteen municipalities which are comprised within the Provinces of Syracuse and Ragusa (Fig. 2). The entire Province of Ragusa is included in the area, while only some parts of the Province of Syracuse are (see PRIN Postmetropoli 2015). The only city with a population over 100,000 inhabitants is Syracuse, with three more towns over 50,000 (Ragusa, Vittoria and Modica) and seven under 10,000.

These municipalities were chosen by applying the criteria of inter-local planning and programming initiatives, i.e. by considering the connections between them in the light of their abilities to team up and play an active role in territorial planning.

The following sections will present some data regarding these two areas. Data are divided into three sections: a socio-economic outlook describing the characteristics of demography, employment and income in Sicily; a spatial analysis of housing and land use; and finally, an institutional description of the administrative state of the art in the two regions. Each section describes the most consolidated trends in this area, but also discusses the most innovative changes and the challenges this area is facing in recent years. The final section points out some concluding remarks and puts the two case studies within the theoretical framework of the post-metropolitan discourse in Italy.

Fig. 2 South-Eastern Sicily (including the Province of Ragusa and part of the Province of Syracuse) (https://www.google.it/maps/)

2 Socio-economic Trends: Time for a Change?

The following section will discuss several data related to the domains of demographics (population and density; housing dispersion index; dependency ratio; foreign citizens; mobility index) and economics (unemployment rates; employees by industry sectors; average per capita income).

The generic outlook of Palermo is closely related to the cliché of a marginal condition, quite far removed from the standards of the high-income, high-quality of life status of Northern Italy. Although it would be foolish to deny such a picture, which is historically rooted in the social, cultural and economic peculiarity of this Mediterranean island, it is also worth noting that this area has been able to express several innovative tendencies in recent decades, and that more particular propensities have been emerging in recent years, as a consequence of the economic crisis and the general deterioration of the socio-economic status in Italy. It is true that the Palermo area is still a deprived one if compared to most of the Northern Italian regions, but the gap appears to be getting somehow smaller and, most importantly, suggests that Palermo might have some hidden resources (still to be fully developed) that could help it find a new place and role in post-metropolitan Italy.

As for South-Eastern Sicily, although its generic outlook might look quite similar to what we have already described for Palermo, things are a bit different here. According to some data (such as location, population and density), SES is even more marginal and peripheral than Palermo, considering its distance from any significant economic centre (with the partial exclusion of Catania, about 60 kms north of Syracuse).

Nonetheless, the socio-economic data we will discuss here describe a richer area than Palermo, with some extremely innovative trends, making SES an exceptional case in all of Italy.

Such a case might probably be compared to other marginal but resourceful areas in Northern or Central Italy, such as the so-called *Chiantishire* around Siena.

2.1 Population Trends

The first set of data we are going to present is related to the demographic domain and starts with the simplest numbers: the ones on population. The population trend in Palermo and its surrounding areas proves quite similar to most other major Italian cities, with the main city initially attracting population and then losing it to its surroundings. This may be considered a traditional case, as with most medium- or big-sized cities in Italy and Western Europe.

Considering the whole Italian state, Palermo has long been the fifth city by population, surpassing Genoa in the 1981–1991 decade (Fig. 3).

The quick and steady growth of Palermo faces a slowdown during the last four decades, but the city is still holding a leading position within its Metropolitan Area, with the second town (Bagheria) only counting a relatively small number of inhabitants, although the ratio between the inhabitants of Palermo and the inhabitants of Bagheria moves from 18:1 (1921) to 12:1 (2011).

Looking at the population percentage variation from 1971 to 2011, most hinterland towns of the Palermo area show very high increases (Isola delle Femmine +176%; Carini +129%), while Palermo is stuck on a mere +2% (Fig. 4).

The Palermo area seems to confirm, at least partially, Edward Soja's (2011) density convergence theory. Although the density of the main city has slightly increased over the forty years between 1971 and 2011 (+91 per sq.km), the surroundings have

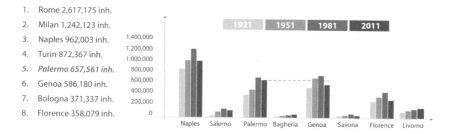

1. Rome 2,617,175 inh.
2. Milan 1,242,123 inh.
3. Naples 962,003 inh.
4. Turin 872,367 inh.
5. *Palermo 657,561 inh.*
6. Genoa 586,180 inh.
7. Bologna 371,337 inh.
8. Florence 358,079 inh.

Fig. 3 Population of Palermo and the second municipality of its area (Bagheria), compared with other similar municipalities in Italy. Time period 1921–2011 (Image by Riccardo Alongi and Giovanna Ceno; data taken from *ISTAT* Census)

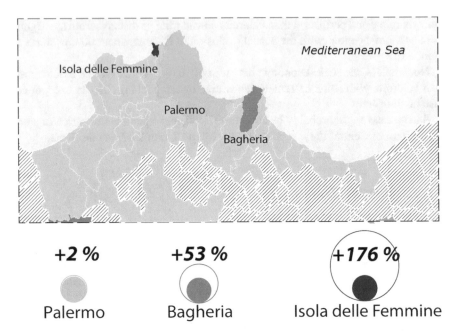

Fig. 4 Percentage variation of population for Palermo, Bagheria and Isola delle Femmine. Time period 1971–2011 (Image by Riccardo Alongi and Giovanna Ceno; data taken from *ISTAT* Census)

experienced a tremendous growth (Villabate +2,490 per sq.km, effectively doubling its original density, to the point of surpassing Palermo as the highest-density town of the area (Fig. 5).

Looking at the population trends, therefore, Palermo is a clear example of centralisation, with a traditional medium-sized city initially attracting population and then stimulating a process of suburbanisation, according to van den Berg et al.'s (1982) urban life cycle theory. Apparently, Palermo is only nowadays reaching a late disurbanisation stage (Picone 2006).

In the case of South-Eastern Sicily, our data show a very unstable attitude here: Modica (the largest town in 1921) slowly yields its leading role to Syracuse, which almost doubled its population between 1951 and 1981. The secondary administrative centre, Ragusa, kept a marginal role in the development of the area (Fig. 6).

Looking at the population percentage variations from 1971 to 2011 and comparing them to other more traditional cases like Palermo, South-Eastern Sicily shows low percentage increases (the highest being for the small town of Acate, +65%) and some significant decreases (like Monterosso Almo, −21%; Fig. 7).

If depopulation can be evoked as an explanation for Monterosso Almo and similar cases, it is harder to explain Acate's growth. We will discuss the reasons for this growth further on, but for now we will just point out that most of Acate's new citizens are foreign citizens.

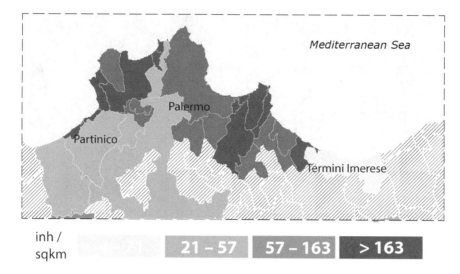

Fig. 5 Density variation for the metropolitan area of Palermo. Time period 1971–2011 (Image by Riccardo Alongi and Giovanna Ceno; data taken from *ISTAT* Census)

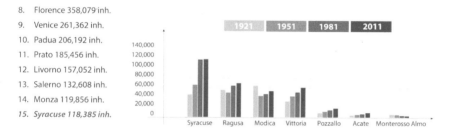

8. Florence 358,079 inh.

9. Venice 261,362 inh.

10. Padua 206,192 inh.

11. Prato 185,456 inh.

12. Livorno 157,052 inh.

13. Salerno 132,608 inh.

14. Monza 119,856 inh.

15. *Syracuse 118,385 inh.*

Fig. 6 Population of Syracuse and other municipalities in South-Eastern Sicily. Time period 1921–2011 (Image by Riccardo Alongi and Giovanna Ceno; data taken from *ISTAT* Census)

When we try to adapt Soja's (2011) density convergence theory to South-Eastern Sicily, we get a surprising result. Unexpectedly, the densest town of SES is the small and only relatively relevant town of Pozzallo (Fig. 8). Even in 1961, Pozzallo was at the top of the list, with a density of 797 inhabitants per sq.km. In 2011, Pozzallo's density (1,231 inhabitants per sq.km) was much higher than Syracuse's (570).

Although this paradox is partially explained by the small size of the municipality of Pozzallo, if we try to apply Soja's density convergence diagram to SES, the results are utterly contradictory. This could probably be considered a nod to the polynuclear system that characterises this region, with at least four greater core areas (Syracuse, Ragusa, Vittoria, Modica) and other smaller districts gravitating around them. Looking at the population trends, the South-Eastern area has long been a place to move away from, and thus does not comply with van den Berg et al.'s (1982)

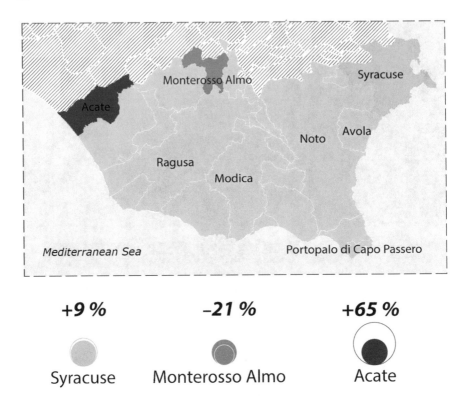

Fig. 7 Percentage variation of population for Syracuse, Monterosso Almo and Acate. Time period 1971–2011 (Image by Riccardo Alongi and Giovanna Ceno; data taken from *ISTAT* Census)

urban life cycle theory. Arguably, most people moved to the nearby industrial cities of Catania and Gela, while others relocated outside of Sicily.

These movements may be due to the rural nature of this area. Only Syracuse thrived, probably as a consequence of its close ties with Catania and the factories of nearby Augusta.

In these last years, however, things have started to change, as SES now hosts several small- or medium-sized towns with increasing growth rates, most of them close to Ragusa. The reasons of this change must be explored within the domain of the recent economic developments, involving a renewed role for agriculture, along with the touristic attractiveness of this region (Picone 2006).

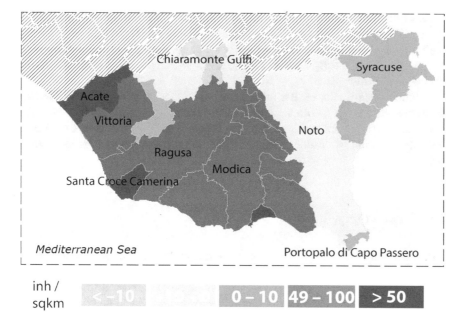

Fig. 8 Density variation for South-Eastern Sicily. Time period 1971–2011 (Image by Riccardo Alongi and Giovanna Ceno; data taken from *ISTAT* Census)

2.2 Housing Dispersion

The analysis of the housing dispersion index (the ratio between the number of scattered houses and the total of houses) in the Palermo area helps to understand how Palermo and its surroundings were behaving in 2001, a transition year between the metropolitan and the post-metropolitan phase.

Moreover, the housing dispersion index is a good measure of how the socio-economic outlook is tightly tied to the spatial one (which will be discussed later on). In the coastal municipalities near the city, just like in Palermo, the dispersion index can be considered low or medium, if compared to other Italian cities (the index hits the 0.05 threshold in 9 municipalities and moves to 0.10 in most of the others).

The farther one moves from Palermo and its immediate surroundings, the higher the indexes become, with a maximum value of 0.30. A single municipality, Bolognetta (in addition to the island of Ustica, which, being an island, cannot be easily compared to inland municipalities, and whose population is mainly tied to the summer tourist presence), has a high dispersion index, greater than 0.40, probably due to its strategic location, adjacent to a major road connecting Palermo to Agrigento and Corleone.

This situation confirms the image of Palermo as a *strong* and compact city. The population probably moves around for working reasons, but there is also a nod to the metropolitan phenomenon, which is not very strong yet, having started in the 1990s in the Palermo area.

Recalling Soja's (2000) theories on the post-metropolis and Bruegmann's (2005) interpretation of the sprawl, we claim that the area of Palermo is quite closely abiding by these models, although the post-metropolitan phase and the disurbanisation phase are not fully developed yet, leaving the city to a late metropolitan and late suburbanisation stage.

South-Eastern Sicily, on the other side, highlights a flattening of the housing dispersion index values, with the lowest values found in Pozzallo and Comiso. This confirms that the traditional landscape is still quite intact in this area, because the low fragmentation is a result of cautious policies at the urban and the agricultural level.

2.3 Dependency Ratio

Continuing our description of the socio-economic outlook of the two regions, we will now consider the dependency ratio, i.e. the ratio of the sum of the number of children (0–14 years old) and elder persons (65 years or over) compared to the working-age population (15–64 years old), as a key factor to understand how both areas are experiencing new trends that might question the legitimacy of some enduring clichés. The dependency ratio could be related to the productiveness and economic strength of the analysed area, although, as a memento, we must not forget the dependency ratio only speculates that the productive part[1] of the population actually has employment but holds no certainties over this trait. Rather, the ratio highlights the presence of large clusters of young (0–14) or old (65+) people depending on the productive parts of society.

In 2011 Palermo had a lower dependency ratio (48) than Milan (60), Rome (58) and Naples (50). Twenty years before, this was exactly the opposite way around, with Milan (39), Rome (38) and Naples (45) all having a lower ratio than Palermo (48). This means that the demographic composition of Milan has changed a lot in those twenty years, while the composition of Palermo is still quite similar to 1991. One likely interpretation of this apparent paradox could lie within the consequences of the economic crisis starting in 2008, which struck the more productive parts of Italy and forced people to relocate to other countries. Sicily, considering its historically marginal role in the productive processes, did not experience a comparable change in its demographic composition. The metropolitan area of Palermo does not show any significant exception to the general trend so far described. In 2011, no single municipality had a particularly higher or lower ratio, and almost all were comprised within the 40–60 range.

These data are even more significant if we compare the highest percentage variation of the dependency ratio from 1991 to 2011 for the municipalities in the Palermo area (Trappeto, +24%) to other Italian cases, like the Milan area (San Donato

---

[1] The dependency ratio is actually just a demographic indicator; linking it to other economic elements (like the unemployment rate or the inactivity rate) would require additional speculation.

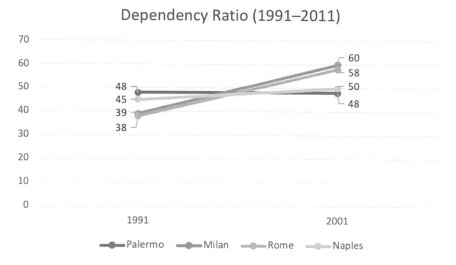

Fig. 9 Dependency ratio index for Milan, Rome and Palermo. Time period 1991–2011 (Image by Marco Picone; data taken from *ISTAT* Census)

Milanese, +124%) or the Venice area (Spinea, +87%). Once again, the demographic composition of Palermo appears steadier, although this is not necessarily a signal of a stable economic status (Fig. 9).

Moving on to South-Eastern Sicily, while the biggest towns of Syracuse and Ragusa have experienced an increase in their dependency ratio (Syracuse moved from 44 in 1991 to 49 in 2011, while Ragusa moved from 49 to 52 in the same period), small municipalities like Acate and Pozzallo had a significant drop in their values: Acate moves from 51 to 44, and Pozzallo from 53 to 46. This means that in 2011 we can find a larger working-age population in Acate than in Milan (60), and that this seems a steady trend (Fig. 10). This phenomenon is linked to several factors: the opening and closing of demographic windows (Golini and Marini 2006), the ageing of the population, and the incoming flows of foreign citizens working in the greenhouses.

These data are even more significant if we compare the highest percentage variation of the dependency ratio from 1991 to 2011 for the municipalities in SES (Portopalo di Capo Passero, +16%) to other Italian cases we already mentioned above. Moreover, most municipalities in SES have negative values in their percentage variation of the dependency ratio, meaning that in 2011 the working-age population was larger than in 1991. Most notable are the cases of Acate (−12%) and Pozzallo (−11%), but 12 out of 19 municipalities in the area had a similar profile.

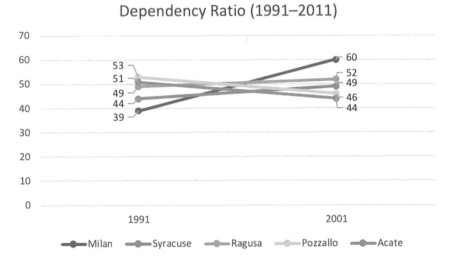

Fig. 10 Dependency ratio index for Milan, Syracuse, Ragusa, Pozzallo and Acate. Time period 1991–2011 (Image by Marco Picone; data taken from *ISTAT* Census)

2.4 Economic Outlook

If we move to more proper economic data and look at the unemployment rate, we must start by considering the general trend for that index in Italy (2001–2019). Generally, the rate was lower in 2011 than it was in 2001, but then it rose and it is now[2] (9.8% in December 2019) higher than it was in 2001, because the economic crisis struck the Italian market a few years later than it did in the US or in other countries. According to the general Italian rate, in most Sicilian towns and cities there was a decrease in the unemployment rate over those ten years, followed by an increase and a new, slow decrease over the following years (the *Provincia*[3] of Palermo moved from 20% in 2004 to 15% in 2011, then to 25% in 2016 and back to 19% in 2018). The unemployment rates for Sicily in 2011 were thus lower than they were in 2004, but still quite high if compared to other Italian regions. In the Palermo area, by considering the data presented for the single municipalities (for the year 2011), it is obvious that unemployment is higher in those areas immediately surrounding the city, like Capaci (31%) and Villabate (29%), and it seems that the farther one moves from Palermo the better it gets (Termini Imerese shows a 22% rate, despite the recent crisis connected to the closing of the *FIAT* factories). Palermo itself has a 23% rate, which is very high if compared to Northern Italian cities (Milan,

---

[2]These data do not take into account the events connected to the Covid-19 crisis.

[3]The *Provincia* (Province) does no longer exist in the Italian administrative system, but it is still used by the National Institute of Statistics (*Istituto Nazionale di Statistica, ISTAT*) for its data. The *Provincia di Palermo* included 82 municipalities and has now been replaced by the *Città Metropolitana di Palermo*, which includes the same municipalities.

Venice and Florence are all around 6%) but in line with other Southern cities (Naples and Catania 26%, Cagliari 17%).

Considering the employees by industry sector, the Palermo region looks quite traditional. Fishing is still quite important in the economic outlook of some towns, but less and less relevant with each passing year (Santa Flavia moved from 40% employees in 2001 in sector A—agriculture and fishing—to a mere 26% in 2011), while agriculture per se is almost worthless and definitely less important, in a strategic sense, than it is elsewhere in Sicily. Even in sector I (which may be linked to tourism), with the exception of Palermo most towns are experiencing negative trends, with very few exceptions. One of the few sectors that still seems quite alive and well is sector G (wholesale trade), but this does not appear like a suggestion that the economic status of Palermo is particularly innovative.

If we refer to Allen J. Scott's (2008) ideas on the most relevant economic aspects in post-metropolises, it seems quite clear that Palermo and its surroundings are quite far from the general trends of global metropolises. Likewise, considering the average per capita income, in the region of Palermo there is a distinct difference between the capital city, where the income of taxpayers is one of the highest of all Sicily, and the rest of the municipalities. This is possibly due to the fact that those who reside in the municipality of Palermo have better-paid jobs, which allow them to live there. The remaining region, generally quite heterogeneous, is divided between coastal areas with a higher income and inland areas with lower incomes.

The average per capita incomes of the inhabitants, if compared to those of the taxpayers, are lower: Palermo has a € 11,073 income per inhabitant, compared to the € 19,867 per taxpayer.

Things are different in Northern Italy, as the difference between the two data is lower (e.g. Venice has a € 17,207 income per inhabitant, compared to the € 22,223 income per taxpayer); this confirms the effects of the high Sicilian unemployment rates, as we have previously discussed.

In the case of South-Eastern Sicily, more surprises come if we look at the unemployment rate. Once again, we can notice the same trend we discussed above for the Province of Syracuse, which went from a 17% unemployment rate in 2004 to a 15% in 2011, then up to a (quite astonishing) 25% in 2015 and finally to a 22% in 2018. The Province of Ragusa boasts even lower unemployment rates: 8% in 2004, 12% in 2011, 19% in 2015 and finally 18% in 2018.

The unemployment rates for Sicily were initially lower in SES, with the municipality of Syracuse having a figure of around 17% in 2011, and Ragusa and Modica getting a respectable (by Sicilian standards) 13%, if we compare it to bigger Southern cities like Naples and Catania (both around 26%). Again, agriculture and tourism may be two leading fields, dictating this more positive (or rather, less negative) trend.

Although the unemployment rate for SES has increased enormously over the last ten years, these latter hypotheses are confirmed if we take a look at the employees by industry sector.

SES proves to be one of the most interesting regions of all Italy, as its Agriculture and Fishing (sector A) percentages of employees are very close to the top of the list (Portopalo di Capo Passero has a remarkable 49%, but most towns in the area have greatly increased their employees in this sector). Acate, one of the smallest towns west of Ragusa, looks like a solid sample of this trend: the employees in agriculture moved from 1.1% in 2001 to 6.3% in 2011, and definitely contributed to the demographic growth of the whole municipality in recent years; much of this growth can be explained with the presence of greenhouses where flowers, tomatoes and eggplants are grown and the increasing presence of foreign people, who are employed in the greenhouses, often as seasonal laborers.

Moreover, wines play an important role, given the growing importance of *IGP* (*Indicazione Geografica Protetta*) and *DOC* (*Denominazione di Origine Controllata*) wines, as the industrial dairy sector does for the same reasons. As for tourism (sector I), most towns in SES have experienced remarkable increases in this domain, particularly Portopalo di Capo Passero (6% in 2001, 13% in 2011) and Pozzallo (7 and 12% respectively). This may be mainly explained by beach tourism, but cultural tourism in the UNESCO cities of the Baroque (Modica, Ragusa, Scicli) also plays a key role (Fusero and Simonetti 2005).

Quality agriculture and tourism are two strategic elements for understanding how SES is slowly but firmly shaping its outlook and changing its representations in the global arena: from a traditional, deprived and marginal periphery to a thriving economic driving force in Sicily, and one to be reckoned with in Italy.

Looking at the most relevant economic aspects of post-metropolises, as Scott (2008) suggests, SES—although incomparable to much bigger and different contexts like Los Angeles—unexpectedly shows some elements that may be defined as post-metropolitan, like a lower unemployment rate and a demographic increase for those small towns (like Acate) hosting a renewed agricultural attractiveness.

Considering the average per capita income per taxpayer, the city of Syracuse (€ 18,026) surpasses Ragusa, Modica, Noto, and Avola with their slightly lower incomes (€ 16,000–14,200). Incomes are also high compared to the rest of Italy. Venice, for example, has an income of € 17,207, lower than Syracuse and higher than Lucca (in Tuscany), where shop rents are more than twice as expensive. The average per capita incomes of inhabitants are lower than those per taxpayer, however: Syracuse has € 11,356, compared to € 18,026 per taxpayer. As we have already remarked, things are different in Northern Italy, because the variance between the two values is lower; this confirms the effects of the higher Sicilian unemployment rates previously discussed, and portrays a society with few, but rich taxpayers.

2.5   Foreign Citizens

All the data we have so far analysed may apparently strengthen the idea of a marginal, deprived land. However, we have already pointed out that some data (like the dependency ratio for Palermo or the unemployment rate for SES) are surprisingly hinting at a less negative situation than the one we would expect. Therefore, portraying Sicily solely as a region without hope is at the very least limiting.

There are clear, if feeble, signs of something evolving both in Palermo and in SES, as slow as it may seem. In order to provide additional details, we now want to return to some other demographic data, and discuss the growing presence of foreign citizens, which is affecting Sicily in unexpected ways (Attili 2008; Lo Piccolo 2013; D'Anneo 2016).

In the twenty years between 1991 and 2011, the Palermo area showed a complex trend, with some municipalities facing negative values (Trabia) and others moving from positive to negative, or the other way around (Santa Flavia, Misilmeri, Villabate and so on). Within this area, most foreign citizens live in Palermo, which proves to be the most attractive city for its employment opportunities, although the percentage of foreign citizens in Palermo in 2011 (around 3%) is still much lower than in Northern or Central Italian cities (Milan 14%, Prato 15%).

These trends lead to a few interesting considerations if we analyse the Index of Dispersion of the Foreign Population (IDFP), an index our research group has built using the existing literature on migrants and foreign citizens (Caritas Migrantes 2011; INEA 2013; Giampino et al. 2014; Lo Piccolo and Todaro 2015), and defined as the percentage of foreign population in a single municipality multiplied by 100 and divided by the percentage of foreign population in the most populous city of that region. In this case, the Palermo area is in line with most of the other Italian regions (e.g. Lombardy, Piedmont), with foreign people living mainly in the most populous city (Fig. 11).

When we look at the data on migration flows in Sicily, we can point out a couple of significant points. The Palermo area is experiencing a slow but steady increase of incoming foreign people, most of them headed towards Palermo; at the same time, the suburbanisation process is causing Italian people living in Palermo to relocate to the hinterland (mainly to Carini and Misilmeri), in search of cheaper housing.

Anyway, the general outlook is quite similar to the trend we can see for Milan and Turin, although these cities show some polarisation phenomena (since 1991 in the case of Turin, or 2011 in the case of Milan).

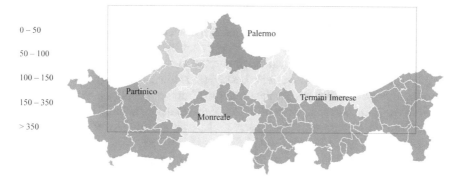

Fig. 11 Index of dispersion of the foreign population for the Palermo area. Time period 2011 (Image by Riccardo Alongi and Giovanna Ceno; data taken from *ISTAT* Census)

0 – 50

50 – 100

100 – 150

150 – 350

> 350

Acate

Ragusa

Santa Croce
Camerina

Syracuse

Noto

Pachino

Fig. 12 Index of dispersion of the foreign population for South-Eastern Sicily. Time period 2011 (Image by Riccardo Alongi and Giovanna Ceno; data taken from *ISTAT* Census)

Looking at South-Eastern Sicily, in the same time frame as discussed above (1991–2011), all the municipalities included in SES have experienced a steady increase of foreign citizens. Acate (19%) and Santa Croce Camerina (15%) display very high values, like most other towns west of Ragusa. Once again, this is tied to the presence of greenhouses and intensive agriculture, where foreign citizens are often employed in deprived work conditions, sometimes approaching slavery and mistreatment.

Most foreign citizens in SES traditionally come from Northern Africa, but recently some Eastern European presences (mainly people coming from Romania) have overcome them.

By analysing the IDFP for this area, Acate (717) and Santa Croce Camerina (581) have the highest index of all Italy, given the relatively low presence of foreign citizens in the most populous city of SES, Syracuse (Fig. 12). This is another hint of the peculiar, polynuclear, post-metropolitan configuration of this region. The Venetian area has some similarities, with a strong polarisation comparable to the westernmost municipalities of SES.

When we look at the data on migration flows in Sicily, we can point out a couple of significant issues. Syracuse is experiencing a slow but steady increase of incoming foreign people, with Italian people moving out of the city and relocating elsewhere (probably due to the high cost of living), while some towns like Ragusa, Modica and Noto all have positive values. At the same time, foreign people are moving to this region in very high numbers, if compared to the original population. This is especially true for Vittoria, Ragusa and Acate. There is also a strong increase of foreign people in Syracuse, but, given the high cost of living in that city, at least a part of these people is likely able to afford that lifestyle, therefore suggesting different national origins (i.e. Western Europeans looking for a historically and culturally attractive region to live in).

2.6   Towards an Unusual Post-metropolitan Region?

This rather quick presentation of the socio-economic data shows that the long-term trends regarding Palermo are intimately related to a marginal condition and a deprived area, as correctly addressed by most scholars who focus on the influence of low employment rates and the destructive presence of criminal organisations (Cannarozzo 2000, 2009; Rossi Doria 2003); however, though these traits are not to be dismissed, there is a serious risk of overestimating them and ignoring the (feeble) traces of something new growing in the background. South-Eastern Sicily, on the other hand, proves to be an exceptional case if compared to other parts of Sicily or Southern Italy. This uniqueness is mainly due to a marginal, yet extremely resourceful status that traces its roots to the Baroque era and creates a space suitable for high-quality tourism and agriculture. In a sense, the most notable path dependence of SES is related to the role it has played within the island, and a somewhat wise exploitation of the traditional resources of the place, combined with a renewed interest for cultural and tourist relationships to other European countries.

The economic crisis of 2008 has somehow hit the Northern Italian regions harder in terms of the relative loss of employment and the worsening of economic parameters, leaving Sicily in a still hindered but possibly more competitive position if compared to other similar Italian cities, and opening new trends that are still uncertain but could prove innovative and unexpected. However, things seem to have been getting relatively worse in the last few years, and the impact of the Covid-19 crisis will require further investigations in the future.

The challenges that Sicily is now facing are first of all connected to its geographic position at the centre of the Mediterranean Sea: given its location, Sicily is quickly turning from a reservoir of emigrants to a crossroads in the often-desperate trajectories that lead immigrants to Europe (Guarrasi 2011). In this very complex context, Palermo may act as a catalyst for promoting new policies of shelter and refuge for migrants: the local municipality has apparently been supporting this approach in the last few years, trying to turn Palermo into not just a geographical, but also a cultural and political hub for those migrants that look at Europe as a promised land.

These processes are strengthening the idea of a post-metropolitan role for the area. As for SES, the relatively small dimension of most cities and towns has likely reinforced the idea of a polynuclear urban region, with no capital centralising functions and policies (as it happens for the case of Palermo). The presence of foreign citizens is an important piece of this puzzle. Therefore, the latest socio-economic developments are changing the traditional image of this apparently peripheral region, possibly turning it into a lesser, yet well-acknowledged cultural and economic polarity in Italy.

3 Spatial Patterns

In Sicily the phenomenon of soil consumption, which started in the 1960s, has strongly contributed to shape the regional territory both from a physical and from a functional point of view. Since then, the soil consumption of the fragile island territory has never entirely stopped. The comprehension of these processes cannot be separated from a close examination of the role that, historically, the building industry and housing revenue have played in the region's economy.

Although the building industry has played a leading role in the economic recovery of the whole of Italy since the 1950s, in Sicily, as in most parts of Southern Italy, this sector has taken on an exceptional importance, as a consequence of the fragility of its productive and social systems.

Since 1960, the data on soil consumption reflect the dispersive model that has characterised many other Italian areas. This model, which is well represented by low-density settlements sometimes supported by illegal practices, is common throughout the whole region. Within the inner areas there are large and unused areas close to agricultural lands. On the contrary, from the 50s to the 90s, coastal areas have been characterised by a totally uncontrolled building growth, becoming a perfect representation of continuous urbanisation (Fig. 13).

Another process, which started in the 80s, has contributed to fill up the coastal area and increase the anthropised land percentage: the suburbanisation phenomenon of the bigger cities on the coast inside the metropolitan areas. This growth headed towards middle cities according to a specific direction, related to the different geographical contexts, and with an increasingly bigger range.

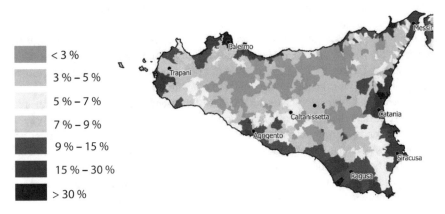

Fig. 13 Soil consumption of the municipalities in Sicily. Rivers, lakes and other water bodies are excluded from the percentage (Munafò 2018, 170)

3.1   Urban Sprawl vs. Polycentric Patterns

The urban structure of the Palermo area has taken on the appearance of the urban sprawl model (Bruegmann 2005). Its principal peculiarities are high soil consumption, high management costs, and significant community flows related to the lack of facilities.

Palermo has a population of 663,401 inhabitants (*ISTAT* data for 2019) and no other municipality of its region boasts a comparable number, with Bagheria (54,714 inhabitants) being the second most populated municipality. To complete the framework of the suburbanisation that marks the metropolitan system we have to talk about the Industrial Development Areas (IDA), grown in the coastal area of Sicily, often localised near areas of natural interest. Moreover, the coastal parts which did not exhibit any industrial areas have been seized by beach establishments or seasonal settlements characterised by a very low density.

The situation of South-Eastern Sicily is entirely different. Spatial phenomena and particularly those linked to the dynamics of the settlement system and other variations of utilising the land, taken in parallel with demographic dynamics and those relating to population distribution, restore a polycentric distribution to SES, in contrast with the accentuated mono-centrism of the greater Metropolitan Area of Palermo. SES comprises the whole Province of Ragusa and the Southern part of the Province of Syracuse (including its administrative centre), where Syracuse, with its 121,171 inhabitants versus the 73,373 inhabitants of Ragusa (*ISTAT* data for 2019), is instrumentally considered the main centre; however the entire urban area is characterised by a polycentric settlement structure, in which small and middle-sized towns are interdependent with respect to the provision of facilities and services (Giampino et al. 2014).

The settlement model of SES is unlike the Palermo area but quite common in the rest of Italy, with a network of small- or medium-sized towns, which have traditionally followed lines of development based on the sharing of higher-ranking services and specialisation. In this case, the small- and medium-sized urban areas are strongly linked to the historic events that assigned specific functions (also related to productive activities) to each of them, establishing interdependent relations among them and with larger urban areas. In particular, in the region of Ragusa, agriculture constitutes the first sector of productive specialisation with relevant results in relation to the innovation of the production, which is recognised at national and international level (Asmundo et al. 2011; Giampino et al. 2014).

3.2   Land Use

The data on the settlements in the Palermo region reflect the dispersive model that has characterised many other Italian coastal areas (Indovina 2003). The areas around the city were initially used for housing reasons, and later on for industrial uses.

Fig. 14 What remains of the *agro* of Palermo today (Photo by Annalisa Giampino)

The coverage ratio of the Palermo region emphasises that over the last decade the urbanisation of Sicily has continued, especially in coastal areas and in those munici-palities that are closer to the capital. This change in the land use has slowly jeopardised the identity of the so-called *agro* of Palermo (Fig. 14).

This process is very different from the one that has characterised the growth of the urban structure of the actual Metropolitan Area of Palermo. In fact, up to the second part of the eighteenth century,

> the *agro* of Palermo, which means the entire coast from Alcamo to Termini Imerese, [...] is characterised by a different legal and institutional organisation. The *agro* is a state-owned land, shared between Palermo and Termini Imerese; it means that there are no barons who have the power to found new cities. These cities, except Trabia, grew with no foundation rules. (Renda 1984, 9)

In the twentieth century, after WW2, there have been various causes of the growth and the rules have also changed as a consequence of new national and local laws. Particularly in the coastal territory, soil consumption is manifest in two different typologies. The first confirms a well-known phenomenon, consolidated in recent decades. The second shows a marked choice towards suburban areas. These phenomena show a metropolitan dimension, particularly for large-scale detail trade, for some manufacturing organisations and for particular types of facilities. Analysing all the consolidated data of the municipalities, the socio-economic gap between Palermo and the other municipalities is ever present.

So, in the wide area dimension and especially with regard to the unfulfilled housing demand, all these considerations confirm the suburban role of the majority of the municipalities. At the same time, it is possible to observe a new plural and post-metropolitan organisation of lifestyle, housing and work (Giampino et al. 2014).

Moreover, this recent organisation has fragmented and restructured the traditional commuting relationship between housing and working, above all because of the still unaccomplished or incomplete metropolitan reality (de Spuches et al. 2002; Picone 2006). This is clear in the case of Terrasini, whose coverage ratio was equal to 22% in 2001, and then to 36% in 2011. Today, these same areas are affected by a strong increase of the presence of medium-sized and larger shopping malls. As for agricultural land use, looking at the maps of agricultural lands between 2000 and 2010, in the region of Palermo there is a clear and pronounced distinction between the Western and the Eastern areas. From Palermo to Balestrate there was a significant increase in the Utilised Agricultural Area (UAA), with the exception of Trappeto (−49%).

On the contrary, in the Eastern area, excluding Santa Flavia and Misilmeri (+88 and +10%, respectively), there is a significant reduction of UAA. Over the years this change of use has been affected by coastal tourism, by new holiday homes with very low density (as we stated earlier), but also by a renewed interest in agriculture (although this is not reflected in a consequential increase of employees in agriculture; Magnaghi 2013), as many scholars have described on a local basis (Cannarozzo 2000; Rossi Doria 2007, 2009; Barbera et al. 2009; Fig. 15).

These reflections are tied to the idea of scattered cities and sprawl and explain these transformations through economic and political reasons rather than cultural outlooks.

On the other side, in the polycentric region of Ragusa the morphology of the territory contributed to the creation of a complex and polycentric settlement scheme where small- and medium-sized urban areas are scattered on the edge of calcarenite terraces opening up towards the coast and creating breath-taking landscapes. Direct relationships can be observed between several settlements in the hills and their equivalent along the coastal strip: Vittoria with Scoglitti; Comiso with Punta Secca (suburb of Santa Croce Camerina); Ragusa with Marina di Ragusa. Most of the residential seasonal growth extends in these coastal areas.

Fig. 15 The renewed interest in agriculture and *orti urbani* (community gardens) in Danisinni, a neighbourhood in Palermo (Photo by Chiara Giubilaro)

Generally, the areas around the largest towns were used initially for housing reasons, and later on for industrial uses and coastal tourism. The settlement coverage ratio in South-Eastern Sicily is very low (although it has a strong impact on the landscape due to settlement dispersion), with the highest percentage being in Pozzallo (22%), and the rest having an average of less than 10%.

The lowest peak is found in the municipalities of Noto and Rosolini, which respectively have a coverage ratio of 1 and 5%, while Syracuse and Ispica exceed the 10% threshold and the rest remains under it. These strong increases, along with the diffusion of greenhouses and an economy gravitating towards agriculture, prove very interesting and meaningful consequences (Fusero and Simonetti 2005; Magnaghi 2013).

3.3 Number of Houses

For years, constant monitoring and several local studies have given the image of a region, Sicily, that suffers from a depopulation of the greater urban centres in favour of the smaller ones, where the population prefers to live mainly (but not only) for economic reasons (Picone 2006).

This aspect could be related to a higher life quality satisfaction in small towns, especially if these are well connected to the main urban centre (*ISTAT* 2013). The data collected for Palermo, analysed year by year and by comparing the whole series, confirm this idea.

This area highlights a change, even if minimal, to the number of houses in the middle ranges, although Palermo obviously maintains a leading position over the remaining municipalities. The difference is still very high. What matters is the confirmation of an increase in houses in neighbouring municipalities, and this is coupled with an increase of population in those municipalities, as a result of relocation for the reasons given above (Fig. 16).

The number of houses, on the other hand, remains quite steady in South-Eastern Sicily. Acate, Santa Croce Camerina and Vittoria, however, have all experienced an increase in their number of houses, and this can be linked (at least for the first two cases) to the demographic growth and the influx of foreign citizens (Fig. 17). Reflections on this indicator agree with what has already been asserted in the cases of the resident population index, the housing dispersion index and the unemployment rate.

3.4 Accommodation Capacity and Tourism

Another interesting set of data, which has a close relationship with the first and with the history of the territory as highlighted before, is the accommodation capacity. These data can be analysed in several ways.

Fig. 16 The increase in the number of houses in the area surrounding Palermo, particularly in the municipality of Trabia (https://www.google.it/maps/)

Fig. 17 The increase in the number of houses in the area of Marina di Ragusa, in South-Eastern Sicily (https://www.google.it/maps/)

First, we can look at them as absolute data, to be linked to the accommodation capacity of each town. Secondly, we can analyse the data comparing all the towns in the areas, estimating the whole accommodation capacity. Moreover, we can study the increase in the considered period. Finally, we can compare our areas with those outside of Sicily.

Matching all the analyses, the overall picture that emerges shows that, in Sicily, the largest city beats all the others. Palermo, in fact, offers a high accommodation capacity, and it is placed at the first level in the analysis. Even if with a big difference at the same level we can find another town, Terrasini, probably due to a large resort complex that hosts a large tourist flow, especially during the summer.

In the other towns, in the considered period, the accommodation capacity shows no significant variation.

However, it is important to highlight that five small towns do not have any hotels, maybe for their location or the low tourist attraction. Another interesting piece of information is that Palermo is located among the top twenty Italian cities for its accommodation capacity.

Recent increases in tourist flows, due to significant cultural events that were organised in Palermo (Arab-Norman Palermo, UNESCO, 2016; Italian Capital of Culture, 2018; Manifesta 12, 2018), prove how the city is strongly committed to the idea of attracting more and more consumers; this process is certainly causing a strong increase in the accommodation capacity (Fig. 18).

In the case of SES, tourism is also playing a key role for the whole area. Field research and scholarly literature explain the phenomenon with the more organised and better touristic offer, also aimed at the international population, that characterised the area in the last decade.

All of this must certainly be related to the Baroque architecture, the growing seaside tourism and the aforementioned imagery of historical landscape that characterises the area. These are elements that the municipalities have enhanced, turning them into a driving force for the economy of the whole area. However, taking a closer look at the phenomena analysed, a few interesting and controversial aspects must be underlined.

At the present time, the current offer is based on a short and fragmented cultural chain, weak in terms of system services and innovative contents if compared to the central role of cultural heritage and colliding with a strong national and international competitiveness in the tourist destinations of cultural interest market.

The Province of Syracuse in Sicily is second only to that of Messina for the number and level of hotels. In 2012, for number of tourists, the Province of Syracuse (1,249,936) comes after those of Messina (3,464,271), Palermo (3,057,733), Trapani (2,084,475), Catania (1,871,849) and Agrigento (1,300,906) (Tourism Observatory data, Department of Tourism, Sport and Entertainment, Regione Siciliana, 2014). The increase in the tourist offer is also certainly related to the inclusion of the "Syracuse and the Rocky Necropolis of Pantalica" site in the "UNESCO World Heritage List" (UNESCO WHL) in 2005.

This popular acknowledgement at local and global level is generally considered a contributing factor to the rise in popularity of the site, in its *appeal* and consequently

Fig. 18 Mega-events like Manifesta 12 are spurring the increase in accommodation capacity in Palermo (Photo by Marco Picone)

in promoting tourism. In the case of Syracuse, the growing tourist offer, together with directly or indirectly linked forms of speculation (mainly related to a distorted vision of promoting tourism development, with serious effects on high-quality soil consumption especially in coastal areas and agricultural landscape), could impair the value of cultural heritage for which the site has been included in the WHL. No significant increase in tourism flows and economy corresponds to the real risk. On the contrary, the process of replacement of traditional handicraft and commercial activities in Ortygia, together with the process of construction on the coastal strip and inland or interventions close to the UNESCO site, shows how the presence of the UNESCO site has been an accelerator for the forms of pressure without the effective promotion of development actions, or—even less so—the implementation of safeguarding actions (Lo Piccolo and Todaro 2014).

In relation to the Ragusa region, if we look at the products, services and facilities for tourists, the region has deeply changed its territorial profile over the last twenty years. Namely two main trends have been recorded: new accommodation facilities have been developed, from hotels and holiday villages only to a wide range of large-, medium- and small-sized facilities, and they are now evenly spread throughout the territory, while in the past they were exclusively located along the coast. Over the last decade, accommodation facilities other than hotels, mainly rural accommodation and

Fig. 19 The "Montalbano effect": tourists are attracted by the TV series and the house of Inspector Montalbano (on the right) actually hosts a B&B (Photo by Giovanna Ceno)

B&Bs, enjoyed a consistent and significant positive trend: in the 2012–2013 period, 206 new facilities came into operation (Lo Piccolo et al. 2015).

Concerning tourist flows, a few surveys carried out in this field (Mantovani 2010; Magazzino and Mantovani 2012), by crossing various basic data, clearly show that in the 2000–2008 period tourist arrivals and overnight stays in the Province of Ragusa, if compared to Sicily, increased by 5.00 and 5.80% with respect to the period 1990–1999, when they were 4.50 and 5.20% respectively.

Moreover, given the broader scenario of the international crisis that is also affecting tourism, in general South-Eastern Sicily seems to maintain positive figures.

According to some scholars, the National Italian Television (*Radio Televisione Italiana, RAI*) series "Inspector Montalbano"[4] (broadcast in Italy and in 18 others countries in the period 1999–2020) significantly contributed to this success (Fig. 19; see also chapter by Todaro et al. for more details on the TV series and its impact on tourism in SES).

Although the "Montalbano effect" has helped the world familiarise itself with this area and has contributed to maintaining high levels of tourism, the actual, relative policies have not been capable of transforming and modernising the quality of tourist facilities and bringing them into line with international standards (Lo Piccolo et al. 2015).

---

[4] At present (12 episodes in the 2012–2015 period) *The Young Montalbano* is broadcast, which deals with the events of the Inspector at a young age.

In fact, there is a very weak, or sometimes even inexistent, system strategy aimed at directing tourist flows and the tourist demand towards a more sustainable, responsible and innovative tourism. In particular, the current offer relies on a short and fragmented supply chain featuring extremely poor innovation with respect to the key role of cultural heritage and system services. Nonetheless, a very strong national and international competitiveness does exist in the market of cultural destinations for tourists.

3.5   Real Estate Market

The real estate market is a valuable source of information for our spatial analysis. The indicator of the average sales price of houses, warehouses, stores, and offices, as well as average lease price of houses, warehouses, stores, and offices is interesting if related to the data on the resident population, the mobility index and the working issues.

Therefore, the high cost of leases, such as those in Palermo and in the coastal municipalities with a higher density, is justified by the decreases of the inner Sicilian areas. The highest ranking city for the average sales price of housing is Palermo, together with Cefalù, whose accommodation is related to seasonal tourism.

Bagheria and Santa Flavia, however, differ in the high price of shops and offices. In this last case an important role is played by the proximity with Bagheria and the functional railway link to other local municipalities and especially Palermo, which is only 17 kms away and part of the same urbanisation.

With regard to industrial warehouses the crisis of the *FIAT* factory has caused obvious consequences. The East Coast, in 2012, was less expensive compared to the West Coast, where Carini and Palermo are the most expensive areas.

In relation to the price of buildings, in South-Eastern Sicily the sales price of houses, warehouses, stores and offices is strongly influenced by the peculiarities of the territory and the economic activities related to them. Syracuse, Modica, Scicli and Ragusa were declared UNESCO sites and are affected by rehabilitation programmes of the historic town, and therefore are the most expensive areas. For stores, rental costs exceed the prices of Palermo, close to the € 2,000 per square meter threshold, and still remain low when compared to many popular tourist destinations of the Italian coast. Just like the region of Palermo, South-Eastern prices decrease as one moves towards the inland.

3.6   Confiscated Buildings

The last data considered concern a specific field that in Sicily, and particularly in the area of Palermo, has played an important role at political level in recent years: the data on confiscated buildings, which must be considered as another aspect of the deregulation that characterises corrupted political systems and hinders public action and the production of *commons* (Donolo 2001; Cremaschi 2009). In fact, the policies of confiscating buildings from criminal organisations represent a different approach to chaotic transformation practices, because it fuels new policies of territorial regulation.

The survey on confiscated buildings, even though it was carried out for only one year, is very interesting both as an indicator for the region and in comparison to the other Italian areas. This topic is closely linked to the areas where mafia, in its various forms, unfortunately has a long and established history. The data show a fair amount of residential and productive buildings that have been confiscated until now.

Like in other cases, Palermo heads the list (Fig. 20), and the reason is easily explicable given the historical origins of the criminal phenomenon in Sicily (Cannarozzo 2000, 2009). In the same high rank, we can find some others towns such as Monreale, Partinico, and Bagheria.

Palermo heads the list even if we consider all the Italian cities, and the ratio between Palermo and Rome, or Palermo and Milan, is 10 to 1 (Palermo has 2,481 confiscated buildings in 2012, whereas Rome has 241 and Milan 210).

The reasons are very clear if we consider the history and the politic relevance of criminal organisations in Sicily. At the same time, though, people started to realise the real weight of mafia in the 1990s with the killings of two judges, Giovanni Falcone and Paolo Borsellino, and this gave way to a new awareness of the phenomenon,

Fig. 20 The association *Centro Studi Paolo e Rita Borsellino* (https://centrostudiborsellino.it, accessed 8 June 2020) is located in a confiscated building in Palermo (Photo by Domenico Giubilaro)

along with a stronger desire to fight it: this is why there are so many confiscated buildings in Palermo today.

Although data on confiscated buildings are not as significant for SES as they are for Palermo, it is a certainty that mafia has regulated the processes of urbanisation and the real estate market (Cannarozzo 2009) in all of Sicily, and we must not forget these phenomena when we discuss the housing issues in Sicily today.

3.7 Post-metropolitan Sicilian Trends

Spatial analysis proves that the two areas we are discussing here each have their peculiar traits. Undoubtedly, in the initial phases the sprawling model that characterised the transformation of identity in the Palermo area was the same model that affected other Italian metropolitan areas.

However, in the most recent years the phenomenon has developed to achieve different outcomes. On the other hand, SES is characterised by a polycentric settlement structure, in which small- and medium-sized towns are interdependent with respect to the provision of facilities and services. This settlement development is based on the sharing of higher-ranking services and specialisation of towns. The polycentric model seems like a more innovative and less traditional spatial pattern for Sicily, one that all of Sicily might probably want to consider for its future development, if given the opportunity.

The actual post-metropolitan vision of the Palermo area is to be connected to a new organisation of lifestyle, housing and work as a consequence of the attempt to start institutionalised metropolitan processes at the beginning of the 1990s, whereas in SES it is related to the international attractiveness of this area, that increased due to the following two factors: marketing activities related to the cultural and gastronomic offer for international tourism and the opportunity to easily find unskilled work in the greenhouses for immigrant workers.

4 Planning a Post-metropolitan Sicily

The institutional choices that have affected the island reveal an ongoing, yet not mature, process, a path which is often associated with existing administrative structures, in terms of aggregations and conservation of roles, positions, and management tools. In this section, we are going to discuss the institutional characteristics of Palermo and South-Eastern Sicily, but one should consider that these two areas have very different social and spatial outlooks, as described in the previous sections. Therefore, even the two institutional contexts will be analysed considering their respective profiles. The main issue concerning the area of Palermo is tied to the institutionalisation of metropolitan cities, whereas SES has a very particular history of inter-institutional cooperation.

4.1   The Creation of a Metropolitan City

Confirming that the institutional process has started but has not taken off yet, Sicily, even before other parts of Italy, has found a renewed interest in the Italian debate on redefining the institutional *metropolitan city* (see chapter by Lotta for additional information). In the 1960s, this interest already existed, when the issues of the city-region and regional planning were debated. The Region[5] had expressed its will to establish a level of intermediate dimension and, with RL 9/1986, instituted the Regional Provinces, aggregations of municipalities into Consortia corresponding to the pre-existing Provinces.[6] The Region had also proposed the identification of metropolitan areas and had defined the criteria for their identification, delimitation, functions and objectives.[7] Since then, the new metropolis was hard to map, as its boundaries became more and more blurred (Picone 2006). The difficulty of that time, like today, was aggravated by the local institutions ignoring the economic territorial features of the affected areas, the spatial and environmental implications, the mobility, the technological innovation, the facility settlement, the offer of services and finally the labour market (Di Leo 1997; Schilleci 2008b).[8]

Recently, the Regional Government has got back to the path of reform to establish the metropolitan cities, which include Palermo. With RL 15/2015, the entire former Province of Palermo is identified as a single metropolitan city. No reasoning about the real dynamics in progress has been carried out. Once again, the institutional model, revealed both in law and in practice, appears to be monocentric. Palermo still wants to maintain a central role, due to its position of political and administrative capital. The criteria to define the new administrative dimensions are not innovative: they are based on income position, in terms of functionality and concentration of certain service categories. Also, there are not many innovations at management level of the new dimension. The Plan for the metropolitan area, dictated by RL 71/1978, should have been managed at municipal or provincial level, because the law did not require an administrative entity of metropolitan dimension. The result was an area with two overlapping plans.

---

[5]The Sicilian Region has a special status, approved by the Constitutional Law of 26 February 1948. This law has regulated the power to legislate on an exclusive basis about certain topics enumerated in the Statute, as local order authorities, urban planning, agriculture and forestry.

[6]The law establishes that the Provinces must adopt the economic and social programme. This will feed into a Plan of Economic Development with social multiannual order to plan and articulate plans, sectoral and territorial projects.

[7]The delimitation of the metropolitan system proposed, including Palermo, was based on administrative (belonging to the same province) and demographic (a population of over 250,000 inhabitants) criteria.

[8]The Metropolitan Area of Palermo which was proposed at that time included 27 municipalities. It was characterised by a core and by a ring made of thick and continuous urbanisation along the coast, between the valleys of the Oreto and Eleuterio rivers. This was later extended from Termini Imerese to Partinico and had a land area of 906 sq.km and a population at 1991 of just over one million inhabitants (1,001,345), equal to 21.15% of the regional population.

In this way, the relationship between the different levels of planning became so confusing and unclear, that the only result was the study for the General Directives of the Inter-municipal Plan for the Metropolitan Area of Palermo, presented in 2001 and including the General Guidelines, regarding cognitive analysis, the Annexes, containing part of the information that were produced, and a Commercial Plan.

Despite the existence of some dynamism, this is not yet supported by a real implementation. Currently, the area of Palermo appears particularly active in promoting local projects.[9] These exceed the average of the largest Italian cities. However, most projects are limited to the Palermo municipality, which has proven unable to include the surrounding municipalities in its development programmes (Giampino et al. 2010).

The partial data available show a strong engagement, but with few results. In fact, the hinterland of Palermo, headed by Monreale, counts two active pacts. With regard to *GAL*,[10] instead, part of the same municipalities belonging to the Territorial Pact are affected by this programme and create an exception in the entire Italian framework. Finally, considering the *PRUSST*, we have the same dynamic and Palermo has 1 *PRUSST* (against the Italian average score of 0.02). Actions for Agenda 21 instead are inactive in the Palermo region.

The absence of a coordination plan for large areas has a negative impact on the territory. Each municipality has planned its territory regardless of the surrounding areas. The largest absence has been and continues to be perceived in the adoption of some complex programmes. These programmes—aiming to enhance the partnership with private stakeholders, to devise new means of propulsion of urban regeneration, etc.—have only affected parts of the territory, with no systemic approach. The programmes have established uncertain relationships within a potential scheme of Palermo's wider area, confirming what was already delineated by the relationships between municipal planning and complex programmes (Lo Piccolo and Schilleci 2005).

The case of complex programming initiatives that affected the eastern part of Palermo's former province leads in fact to reflect about an incomplete or inconsistent dialectic between the projects expressed by the various tools and the complex programmes. The latter sometimes show elements of a post-metropolitan innovation for the local territorial realities, looking for a difficult coordination with the municipal planning and with the (unrealised) wide-area planning.

---

[9]Since the beginning of the 1990s, "the new instruments, known as local development partnership programmes, were conceived by the ministry to support the development and implementation of specific projects through cooperation between the public and private sectors" (Lo Piccolo and Schilleci 2005, 80). The scene quickly becomes much larger, with the introduction of the so-called "complex urban programmes" as *Programmi Integrati di intervento (PII), Programmi di Recupero Urbano (PRU), Programmi di Riqualificazione Urbana (PRIU), Programmi di Riqualificazione Urbana e Sviluppo Sostenibile (PRUSST)*. These are consolidated in parallel with the experiences due within the framework of EU programmes, like Agenda 21.

[10]The *Gruppo di Azione Locale (GAL)* is a local action group composed of public and private stakeholders to promote local development in a rural area. *GAL* are funded by the EU Initiative Programme called LEADER +.

Indeed, Palermo as a metropolitan city continues to maintain its rank, its role, its attractiveness and innovation, but at the same time a few surrounding towns, albeit slowly, begin to structure a potential system of polynuclear city-region. In continuation of the paths taken in the 1990s,[11] some local governments have in fact had the power to promote a territorial coalition among the municipalities that fall between the valleys of Imera Settentrionale and Torto, the Madonie park authority, the former Province of Palermo and a public–private partnership.[12]

This coalition has created some forms of coordination.[13] These actions were coordinated only due to the will of the promoters and managers and are not included in a wider range of vision, but the most recent results are an indicator of an uphill process, characterised by a regional and introverted vision, that basically ignores the flows and dynamics that cross it.

Some municipalities within the Metropolitan City of Palermo, such as Termini Imerese, have expressed their intent to continue a difficult, but not unsubstantiated, path towards polycentrism, by working together with other nearby municipalities in a common framework. Under national influences, in 2013, the regional government got back to the idea of creating the metropolitan cities. RL 8/2014 established three metropolitan cities, including Palermo, and despite the backwardness of the criteria they proposed (once again, these were territorial continuity and population) there is an element of innovation: RL 15/2015, in article 45, paves the way for the establishment of new consortia.

In those places where we find the result of a decentralisation or re-centring, a deterritorialisation or reterritorialisation, a continuous extension or urban nucleation intensified, a growth of the homogeneity and heterogeneity, a socio-spatial integration

---

[11] At the time Termini Imerese, Terrasini, Trabia, Capaci and Palermo appealed to the Regional Administrative Court (*Tribunale Amministrativo Regionale, TAR*) asking for the revocation of the institutive decree. The reasons for this opposition were of a different kind, but all aimed at soliciting the Regional Government to revise RL 9/1986 so that better account was taken of the indications of the NL. One of the main demands considered the establishment of an elective organ of government for the metropolitan area (the metropolitan city) and the redefinition of the concept of metropolitan area more as a *system of cities*, as configured in the NL, than as an area centred around a capital city, as suggested by the RL. A short step to redefine the Metropolitan Area of Palermo was made, identifying not only administrative boundaries, but trying to work on local systems greatly consolidated and based on the elements and environmental reports as fundamental to this delimitation. The focus was also placed on the need for integration between the metropolitan system and regional territory, as well as on internal relations. These reflections, however, never followed through and did not resolve either the problem of delimitation or, even less, the big deal of the liaison with the provincial and municipal planning.

[12] The latter is made up of the development company *SOSVIMA* (*Agenzia di Sviluppo locale delle Madonie*) (as technical coordinator) and *IMERA SVILUPPO*, the *GAL Madonie*, the *Banca di Credito Cooperativo San Giuseppe*, the *Banca di Credito Cooperativo Mutuo Soccorso*, *Conf-cooperative Unione Provinciale* of Palermo, *Fare Ambiente Coordinamento Regionale Sicilia*, *Confederazione Italiana Agricoltori Palermo* and *Confindustria Palermo*.

[13] The *Distretto Culturale delle Madonie* (2007), the *Distretto delle carni bovine delle aree interne della Sicilia* (2007), the *GAL Madonie* (2010), the *Distretto turistico di Cefalù e Parchi delle Madonie e di Himera* (2011), the *Gruppo Azione Costiera Golfo di Termini Imerese* (2013) and the *PIST (Piano Integrato Sviluppo Territoriale) Città a Rete Madonie-Termini* (2009).

and disintegration, where perhaps we can already talk about a new geography of post-metropolitan urbanisation (Soja 2000).

In this case, the institution of the Metropolitan City of Palermo could provide an opportunity to delineate the creation of a strong and stable metropolitan government, able to perform functions already defined in the 1980s, while respecting processes undertaken by individual local realities.

Despite the absence of approved coordination plans, the draft plans, such as the *Schema di Massima* and the *Quadro Propositivo con Valenza Strategica* (*QPS*), reveal a polycentric territorial and organisational will.

The reasons for the great interest for polycentrism reside into at least two specific aspects: the transport systems improvements and the introduction of new production methods and new lifestyles, previously concentrated only in the space of the capital city.

The planning of the metropolitan dimension of Palermo would be a great and necessary change of scale in the territorial organisation of the entire coastline, where the greatest flows and exchanges of territory are concentrated.

A thorough planning which is aware of the residential issues, in a complex space like Palermo city, could define the relationship between the different functions and specific territories. Implementing initiatives of coordination, advocating and supporting local economic systems, could in fact start to densify the dynamic connections that are somehow reticular, in terms of interdependence and complementarity, and can structure and define the different parts of the metropolitan area being set up.

Another field of metropolitan area planning should also take environmental issues into account. In this regard, the draft of the Ecological Network could act as a focal point as well as the reorganisation of the waste disposal system, too often oblivious to its territorial impact. Finally, the re-use in a new virtuous cycle of the buildings that are confiscated from mafia would represent an aspect that is not only civic and/or symbolic, but ultimately relevant, even in quantitative terms, within the territory under examination.

4.2 A Fragmented Territory, with Strong Connections

South-Eastern Sicily, the second case study of this chapter, holds no metropolitan city. However, the area shows a high level of local and inter-local planning and programming initiatives. From the planning point of view, the Provinces of Ragusa and Syracuse are traditionally characterised by a greater number of planning instruments than the other provinces in Sicily, dealing with aspects of both territorial planning and environmental and landscape safeguard.

Furthermore, more recently the area is characterised by inter-institutional cooperation practices in order to promote new socio-economic development programmes (*Progetti Integrati Territoriali, Patti Territoriali, Progetti Integrati*

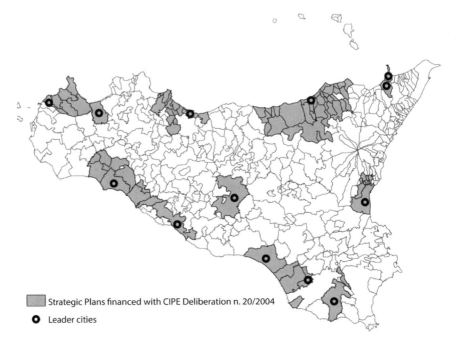

Fig. 21 Strategic plans in Sicily, year 2004 (Di Giacomo 2010, 40)

*d'Area, Programmi di Riqualificazione Urbana e Sviluppo Sostenibile del Terri-
torio, Programmi di Recupero Urbano, Programmi di Iniziativa Comunitaria, Piani
Strategici, Piani Integrati di Sviluppo Urbano, Piani Integrati di Sviluppo Territoriale*). However, these tools establish controversial relationships with respect to the
traditional urban planning instruments and policies (Lo Piccolo and Todaro 2014;
Figs. 21–22).

The new instruments have considerable financial resources at their disposal. In
contrast, town-planning policies are essentially perceived as regulative or, even
worse, as restrictive. In many cases a real clash between the former and the latter can
be perceived.

On the one hand, local development policies have distributed considerable financial resources in a context of fiscal crisis and serious economic deficiency in local
administrations, and have therefore imposed themselves with the *supremacy of
money*.

On the other hand, traditional town-planning policies have not been understood by
local communities and authorities as real opportunities for guiding and stimulating
local development and have often been put into practice in a bureaucratic manner
(Lo Piccolo and Schilleci 2005; Lo Piccolo and Todaro 2014).

However, the results emerging from the new instruments often prove to be
short-lived and incapable of activating effective and long-lasting processes of
socio-economic development.

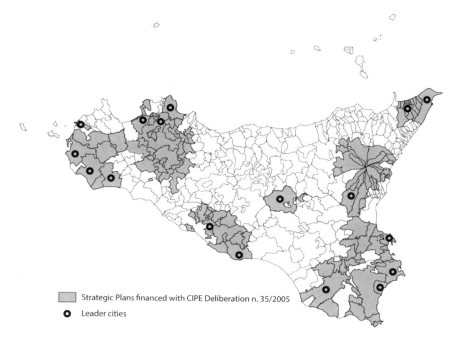

Fig. 22 Strategic plans in Sicily, year 2005 (Di Giacomo 2010, 40)

The general absence of any link-up with town-planning should also be emphasised. In fact, also in cases in which the new instruments are administratively handled by the town-planning sector of the same municipality, they assume the form of isolated projects in most of the cases, with the result of generalised town-planning provisions that are generally feeble.

Moreover, the paradoxical result is that the new instruments, rather than playing a *leading role* in promoting innovative strategies and actions, become a *collection* of goals and actions deriving from other pre-existing programming instruments.

With regard to the topics examined, the main elements of continuity with the recent past seem to consist in singling out tourism as a lever in activating processes of local development and enhancement. It has to be said that these objectives have already been widely associated with integrated territorial planning (which has incorporated many of the new instruments), but these new instruments tend to be more specialised with regard to the sub-sections of cultural tourism.

Although these instruments are characterised by a bottom-up approach, typical of a post-metropolitan reality, the result is an *inverse*, and totally inefficient conformity, which is that of the new instruments compared to the other existing local programmes (Lo Piccolo et al. 2012).

With reference to partnerships and territorial aggregations regarding these local policies, the need to build networks, usually encouraged by National and Regional guidelines, finds an atypical variant in South-Eastern Sicily.

We can identify two main territorial nodes: the urban area of Syracuse and the territory of Ragusa. In the first case, although active programmes have aggregated several municipalities (Augusta, Noto, Avola, Syracuse) in different forms from time to time, territorial coordination processes were activated (e.g. *Piani Strategici*) to bring individual projects to common strategies.

However, the attempt to establish a coalition proves a formal aggregation of municipalities, unable to work for common aims. Anyway, Syracuse remains a reference city for local development policies and tends to activate autonomous programmes.

In the second case, local development strategies describe two main areas of territorial aggregation of municipalities: the mountainous area of the system of Hyblaean Mounts (Giarratana, Monterosso Almo, Comiso, Modica, Ragusa) and the coastal system (Acate, Vittoria, Santa Croce Camerina, Scicli, Pozzallo) (Giampino et al. 2010).

Also, in this case the aggregations of municipalities often produce weak alliances geared towards competitiveness in the accumulation of partners, in order to obtain public funding (especially structural funds), rather than effectiveness of strategy for territorial growth and development. This status appears evident from the fitful commitment of mayors and town councillors in launching new programmes (strategic plans).

In some cases, the role of programme coordination is entrusted to territorial development agencies, which were set up during the running of previous programmes, like, for example, offices (urban centres, civic centres, etc.) for EU policies or European projects spawned in the larger cities.

We can observe a tangible example of this condition in Ortygia, Syracuse's old town. Looking specifically at the outcomes of these instruments in the case of Ortygia, an urban context significantly affected by changes in physical, economic, social and environmental components emerges. However, some critical issues can be observed (Lo Piccolo and Todaro 2014).

In recent years, several projects of urban regeneration have been enabled, with incentives for creating renovation and economic revitalisation initiatives. The Ortygia peninsula was then involved in a process of renewal that, also through the localisation of key administrative functions, has reinstated centrality to the peninsula that has returned to be inhabited by the Syracusans (Liistro 2008). This process resulted in the reactivation of the housing market, also thanks to the significant presence of foreigner investments, significantly increasing real estate values (Cannarozzo 2006).

Although in the last twenty years the peninsula of Ortygia has attracted the interest of planning and programming, the activated instruments have resulted in the realisation of (sometimes only partial) *punctual* interventions that did not follow a unitary and organic project (Lo Piccolo and Schilleci 2005; Lo Piccolo 2007). It shows, in fact, the evident imbalance between the localisation of recovery interventions carried out in the proximity of archaeological and historic monumental interest areas or the seafront and inland areas, which are still characterised by the serious condition of physical and social degradation.

Added to this is the growing investment by individuals and real estate companies, not governed by public action, which determines the progressive replacement of traditional socio-economic network with commercial and tourism activities. This phenomenon, accompanied by the progressive disappearance of neighbourhood services (especially for children and the elderly), is causing the loss of the minimum requirements to ensure habitability (Lo Piccolo and Todaro 2014).

Furthermore, as regards policies for enhancing the cultural heritage in South-Eastern Sicily, a specific phenomenon has been observed, with particular importance for the significant and often controversial effects it produced. This phenomenon stems from a process promoted by the Agencies for Cultural and Environmental Heritage of Syracuse and Catania, later joined by the city of Ragusa, and consists of the *construction* of the unitary territorial image of the *Late Baroque Cities.*

Such image of the territory, based on the recognition of Baroque architecture and urban-planning as a unifying identity value attracting tourists and visitors, inspired the cultural enhancement policies that were implemented in the 1990–2010 period. Among them, in 2002, the "Late Baroque Cities of Val di Noto" (Noto, Scicli, Ragusa, Militello Val di Catania, Caltagirone, Palazzolo Acreide, Catania, Modica) were listed as "UNESCO World Heritage Sites", and the Southeast Cultural District "Late Baroque Cities of Val di Noto" (financed in 2009 by the Regional Operational Plan of Sicily 2000–2006, Measure 2.02.d) was established with the purpose of implementing the UNESCO site management plan (Lo Piccolo et al. 2015). Moreover, the image of this area, produced by the TV series and in particular "Inspector Montalbano" (see Sect. 3.4), also contributed to the construction of this phenomenon.

In the light of this phenomenon, and of the misrepresentation that inevitably comes with it, it should be however pointed out that tourism is a recently emerged opportunity for the territory of Ragusa, which needs well-structured strategies and enhanced consolidation (Trigilia 2012; Azzolina et al. 2012).

In this territory, the tourist districts were initially established with a spirit of cooperation and they are currently nothing but an aggregation of municipalities unable to express a "unitary vision of the territory of the Southeast" (Azzolina et al. 2012, 161).

Moreover, the supply of tourist services is based on a traditional model of tourism, which is fully focused on accommodation facilities and catering services. Such a model features extremely poor innovation with respect to the key role of cultural heritage, despite a very strong national and international competitiveness in the market of cultural destinations for tourists. In particular, services and infrastructures (including technological ones) prove to be inadequate both in the private and public sector (Lo Piccolo et al. 2015).

4.3 Two Different Ways of Being Post-metropolitan

With regard to the dynamics of post-metropolitan institutional prospects, the collected data and the elaborations of our research suggest that the Palermo area has never been fully metropolitan either in its institutional structure and its government, or in its socio-economic and territorial features. In recent years, however, the data analysed in this research show the will to propel the territory into a polynuclear dimension (which somehow characterises the post-metropolitan dimension), with all ambiguities and contradictions of the path. Particularly, the components of change and innovation seem to emerge from some geographical areas which are less dependent on role, geography and functions from Palermo's municipality. They contrast and revolutionise the inertia of the centripetal administrative and hierarchically predominant dimension, which has so far prevented and delayed, in terms of choices and effectiveness, a real metropolitan governance characterising the post-metropolis. The continuous reconsideration of the form and contents of the new metropolitan structure, fluctuating between the metropolitan city model and the consortia, suggests that these innovative elements can actually trigger a process of (post-)metropolisation, in a territory that is actually quite fitted for these changes.

South-Eastern Sicily, on the other hand, shows a high level of local and inter-local planning tools. From the planning point of view, the Provinces of Ragusa and Syracuse are traditionally characterised by a greater number of planning instruments than the other provinces in Sicily, dealing with aspects of both territorial planning and environmental and landscape safeguarding. More recently the area shows a marked tendency to inter-institutional cooperation, by activating new territorial development programmes. These recent experiences are interesting for two reasons: the capability to build networks of inter-institutional cooperation, usually encouraged by National and Regional guidelines and the presumed *flexibility* of these programmes with respect to the *rigidity* of the traditional planning instruments. New instruments often become a *collection* of goals and actions deriving from other pre-existing programming instruments rather than playing a *leading role* in promoting innovative strategies and actions; new programmes also establish controversial relationships with respect to the traditional urban planning instruments. Furthermore, the networks of inter-institutional cooperation often produce weak alliances geared towards competitiveness in the accumulation of partners, in order to obtain public funding, rather than effectiveness of strategy for territorial growth and development.

5   Conclusions

In this chapter, we started discussing the Metropolitan Area of Palermo by recalling how important marginality and isolation are to grasp the Sicilian situation. Part of the data we have discussed so far implies that this marginality causes a lot of social, economic and political issues. For instance, the number of confiscated buildings is proof of the failure of the policies led by the national state and the local authorities, together with a sort of quiet resignation to an economic and social negative status that might seem to confirm the clichés we introduced at the beginning of our analysis.

However, other data suggest a different approach. We have discussed how the dependency ratio of Palermo is now unexpectedly lower than Milan's. This does not mean that finding a job in Palermo is easier than it is in Milan, of course, but it implies that Palermo and its region have some *potential energy* (given by the relatively high number of working-age population). This sort of energy may be the same that causes so many confiscated buildings to serve a renewed purpose, often for social and cultural goals. We could not discuss the use of confiscated buildings more thoroughly for a lack of data, and most reflections on that topic should use qualitative data instead of merely quantitative data; nonetheless, the number of social bottom-up proposals coming from young and unemployed people has definitely grown in the last decade in Palermo, and may be considered a sign of a slow but unyielding investment on social awareness and bottom-up policies.

Even if we consider all of these new trends and potentialities, there is still no simple answer to the question about the post-metropolitan status of this area. How does Palermo relate to the urban regionalisation processes we have previously described (see chapter by Lo Piccolo et al.)? If we compare this city to Milan, Rome, or Los Angeles, there are indeed many differences, but also a few striking elements that should be taken into account. Considering the presence of foreign people or the density convergence theory, Palermo is behaving almost exactly like Milan or Rome, though the numbers are obviously lower in comparison. In our opinion, this means that a portrait of Palermo and its area must be carefully balanced on a tight line, hanging in the balance between the cliché of a marginal and deprived Southern city and the acknowledgement of something new that might come in the future. In a sense, Palermo is arguably experiencing the initial stages of an urban regionalisation process.

South-Eastern Sicily, on the other hand, is probably yet more peculiar. Even temporarily neglecting any social, urban or economic analysis, it is enough to take a simple look at the demographic profile of Los Angeles on one hand and of Palermo on the other hand to understand how those scales cannot be superposed at all. Even worse if we take into account the area between Syracuse and Ragusa. And yet, Soja argues:

> the grounding of the postmetropolitan transition in Los Angeles is not meant to constrict interpretation of the postmetropolis just to this singular and often highly exceptional city-region. Rather, it is guided by an attempt to emphasize what might be called its generalizable particularities, the degree to which one can use the specific case of Los Angeles to learn more about the new urbanisation processes that are affecting, with varying degrees of intensity, all other cityspaces in the world. (Soja 2000, 154)

In other words, if we want to *test* the existence and the possible functioning of post-metropolitan systems, it is not a matter of adapting the Los Angeles model to the world, but of extrapolating from the particular Californian case those *lessons* that can be valid for all the world. The same thing could be repeated for urban regionalisation processes.

It is a paradoxical game, of course, based on the *what if* rule, as *counterfactual history* suggests (Ferguson 1999): in that kind of history, every essential question begins with *What if...?* We believe that in addition to the counterfactual history we might think in terms of a counterfactual geography. Instead of asking ourselves, like historians would do, what would have happened if Hitler had won the war, we will ask what would happen if South-Eastern Sicily were a post-metropolitan land. Our goal is to ascertain whether SES can show to the world some variations to the standard urban regionalisation model.

There are two basic reasons that can help us further this reasoning: first, if we look at the number of employees by industry sectors, within the boundaries imposed by the ongoing crisis, there are interesting variations that help us to outline a more post-metropolitan territorial profile than the one of Palermo, where the traditional leading sectors, such as building manufactures, remain the same.

Second, if we look at the effects that the economic transition raises on spatial structures, we can detect in South-Eastern Sicily a polynuclear localisation process of productive, industrial and non-industrial activities, which follows the historic poly-centric settlement pattern, compared to a conversion of the industries to commercial activities already affected by a process of delocalisation of the centre of Palermo. Similarly, in reference to the ability of internationalisation of the agricultural products of South-Eastern Sicily, this area proves capable of innovating its productive district (Asmundo et al. 2011).

If the area of Palermo, from a normative and conceptual point of view, can at least be considered a metropolitan city, South-Eastern Sicily, traditionally considered a non-metropolitan context, shows a more dynamic, innovative and post-modern situation. What if, as a consequence, South-Eastern Sicily were a new reality able to provide useful insights on possible future alternative urban regionalisation models? And what if, paradoxically, South-Eastern Sicily was even more post-metropolitan, in some respects, than Los Angeles?

One of the topics that seem particularly innovative in SES is the way planning has affected these areas, because most municipalities in the Provinces of Ragusa and Syracuse, as we recalled earlier, produced lots of urban and territorial plans, many more than the other provinces in Sicily did. These tools, however, should interact with the socio-economic policies starting from the challenges that we have described (migrants, cultural tourism, quality agriculture, etc.). If these two domains (planning and socio-economic policies) are able to properly interact, SES might launch development processes that are more efficient than the average of Southern Italy, once again proving to be a leading region in this part of the country.

References

Abbate G (2011) La valorizzazione dei centri minori come elemento strategico dello sviluppo del territorio. In: Toppetti F (ed) Paesaggi e città storica. Teorie e politiche del progetto. Alinea, Florence, pp 141–144

Asmundo A, Asso PF, Pitti G (2011) Innovare in Sicilia durante la crisi: un aggiornamento di Remare controcorrente. StrumentiRes 3(4):1–7

Asso PF, Trigilia C (eds) (2010) Remare controcorrente. Imprese e territori dell'innovazione in Sicilia. Donzelli, Rome

Attili G (2008) Rappresentare la città dei migranti. Jaca Book, Milan

Azzolina L, Biagiotti A, Colloca C, Giambalvo M, Giunta R, Lucido S, Manzo C, Rizza S (2012) I beni culturali e ambientali. Ragusa. In: Casavola P, Trigilia C (eds) La nuova occasione. Città e valorizzazione delle risorse locali. Donzelli, Rome, pp 151–162

Barbera G, La Mantia T, Rühl J (2009) La Conca d'Oro: trasformazione di un paesaggio agrario e riflessi sulla sostenibilità. In: Leone M, Lo Piccolo F, Schilleci F (eds) Il paesaggio agricolo nella Conca d'Oro di Palermo. Alinea, Florence, pp 69–95

Bruegmann R (2005) Sprawl: a compact history. University of Chicago Press, Chicago

Cannarozzo T (2000) Palermo: le trasformazioni di mezzo secolo. Archivio di Studi Urbani e Regionali 67:101–139

Cannarozzo T (2006) Dal piano ai progetti. Due interventi pubblici di recepero residenziale a Ortigia. In: Trapani F (ed) Urbacost. Un progetto pilota per la Sicilia centrale. FrancoAngeli, Milan, pp 194–201

Cannarozzo T (2009) La governance mafiosa e l'assalto al territorio. In: Leone M, Lo Piccolo F, Schilleci F (eds) Il paesaggio agricolo nella Conca d'Oro di Palermo. Alinea, Florence, pp 39–51

Cannarozzo T (2010) Centri storici e città contemporanea: dinamiche e politiche. In: Abbate G, Cannarozzo T, Trombino G (eds) Centri storici e territorio. Il caso di Scicli. Alinea, Florence, pp 9–22

Cremaschi M (2009) Il territorio delle organizzazioni criminali. Territorio 49:115–118

D'Anneo G (2016) Abbandonare o scegliere Palermo, dalla de-urbanizzazione alle nuove migrazioni. http://www.strumentires.com/index.php?option=com_content&view=article&id= 627:abbandonare-o-scegliere-palermo-dalla-de-urbanizzazione-alle-nuove-migrazioni&catid= 16:immigrazione&Itemid=140. Accessed 31 May 2020

de Spuches G, Guarrasi V, Picone M (2002) La città incompleta. Palumbo, Palermo

Di Giacomo G (2010) L'accompagnamento della pianificazione strategica in Sicilia. In: Vinci I (ed) Pianificazione strategica in contesti fragili. Alinea, Florence, pp 29–42

Di Leo P (1997) Area metropolitana di Palermo. Città e Territorio, Bollettino del Dipartimento della Città e Territorio dell'Università di Palermo 3:72–79

Di Leo P, Esposito G (1991) Palermo: Pianificazione urbana e Metropolitana. Due studi. In: Atti del Seminario dell'INU, Palermo, 28–29 January 1991

Donolo C (2001) Disordine. L'economia criminale e le strategie della sfiducia. Donzelli, Rome

Ferguson N (1999) Virtual history: towards a 'chaotic' theory of the past. In: Ferguson N (ed) Virtual History. Basic Books, London, pp 1–90

Fusero P, Simonetti F (eds) (2005) Il sistema ibleo. Interventi e strategie. Idealprint, Modica

Giampino A, Todaro V, Vinci I (2010) I piani strategici siciliani: interpretazioni di territorio ed orientamenti progettuali. In: Vinci I (ed) Pianificazione strategica in contesti fragili. Alinea, Florence, pp 43–93

Giampino A, Picone M, Todaro V (2014) Postmetropoli in contesti al margine. Planum 2(29):1308–1316

Golini A, Marini C (2006) Aspetti nazionali ed internazionali delle popolazioni considerate da una finestra demografica. Quaderni di ricerca. http://docs.dises.univpm.it/web/quaderni/pdf/spec/002.pdf. Accessed 31 May 2020

Grasso A (1994) Le Aree metropolitane siciliane. Funzioni, vincoli, strategie. Pàtron, Bologna

Guarrasi V (2011) La città cosmopolita. Geografie dell'ascolto. Palumbo, Palermo

Indovina F (2003) È necessario diramare la città diffusa? Le conseguenze sul governo del territorio di un chiarimento terminologico. In: Bertuglia CS, Stanghellini A, Staricco L (eds) La diffusione urbana: tendenze attuali, scenari futuri. FrancoAngeli, Milan, pp 116–131

INEA (2013) Indagine sull'impiego degli immigrati in agricoltura in Italia 2011. INEA, Rome

ISTAT (2013) Urbes. Il benessere equo e sostenibile nelle città. http://www.istat.it/it/files/2013/06/Urbes_2013.pdf. Accessed 31 May 2020

Liistro M (2008) Ortigia: memoria e futuro. Edizioni Kappa, Rome

Lo Piccolo F (2007) Siracusa: misconoscimento e potenzialità dell'identità locale. In: Rossi Doria B (ed) Sicilia terra di città. Istituto Geografico Militare, Florence, pp 150–175

Lo Piccolo F (2009) Territori agricoli a latitudini meridiane: residui marginali o risorse identitarie? In: Lo Piccolo F (ed) Progettare le identità del territorio. Alinea, Florence, pp 11–42

Lo Piccolo F (2013) Nuovi abitanti e diritto alla città: riposizionamenti teorici e responsabilità operative della disciplina urbanistica. In: Lo Piccolo F (ed) Nuovi abitanti e diritto alla città. Un viaggio in Italia. Altralinea, Florence, pp 15–32

Lo Piccolo F, Schilleci F (2005) Local development partnership programmes in Sicily: planning cities without plans. Plann Pract Res 20(1):79–87

Lo Piccolo F, Todaro V (2014) From planning to management of cultural heritage sites: controversies and conflicts between UNESCO WHL management plans and local spatial planning in South-Eastern Sicily. Eur Spat Res Policy 21(2):47–76

Lo Piccolo F, Todaro V (2015) Concentración vs dispersión de los inmigrantes en Italia. Análisis comparativo de la distribución de la población extranjera en las regiones urbanas. CyTET 47(184):397–404

Lo Piccolo F, Leone D, Pizzuto P (2012) The (Controversial) Role of the UNESCO WHL Management Plans in Promoting Sustainable Tourism Development. J Policy Res Tourism, Leisure Events 4(3):249–276

Lo Piccolo F, Picone M, Schilleci F (2013) Forme di territori post-metropolitani siciliani: un contesto al margine. Planum 27(2):46–51

Lo Piccolo F, Giampino A, Todaro V (2015) The power of fiction in times of crisis: movie-tourism and heritage planning in Montalbano's places. In: Gospodini A (ed) Proceedings of the International Conference on Changing Cities II: Spatial, Design, Landscape Socio economic Dimensions. Porto Heli, Peloponnese, Greece, 22–26 June 2015, Grafima Publ, Thessaloniki, pp 283–292

Lotta F (2015) Palermo. In: De Luca G, Moccia D (eds) Immagini di territori metropolitani. INU Edizioni, Rome, pp 114–119

Magazzino M, Mantovani M (2012) L'impatto delle produzioni cinematografiche sul turismo. Il caso de il Commissario Montalbano per la Provincia di Ragusa. Rivista di Scienze del Turismo 1:29–42

Magnaghi A (2013) Riterritorializzare il mondo. Scienze del Territorio 1:31–42

Mantovani M (2010) Produzioni cinematografiche e turismo: le politiche pubbliche per la localizzazione cinematografica. Rivista di Scienze del Turismo 3:8–103

Migrantes C (2011) Dossier statistico immigrazione 2011. XXI Rapporto. IDOS Edizioni, Rome

Munafò M (ed) (2018) Rapporto ISPRA. Consumo di suolo, dinamiche territoriali e servizi ecosistemici. ISPRA, Rome

Nobile MR (1990) Architettura religiosa negli Iblei. Dal Rinascimento al Barocco. Ediprint, Syracuse

Picone M (2006) Il ciclo di vita urbano in Sicilia. Rivista Geografica Italiana 113:129–146

Pinzello I (ed) (2003) Il ruolo delle aree metropolitane costiere del Mediterraneo. Alinea, Florence

Piraino A (1988) Il sistema metropolitano di Palermo. Quale fisionomia. Celup, Palermo

PRIN Postmetropoli (2015) Atlante web dei territori postmetropolitani [web atlas]. http://www.postmetropoli.it/atlante. Accessed 31 May 2020

Renda F (1984) Le borgate nella storia di Palermo. In: Ajroldi C (ed) Le borgate di Palermo. S. Sciascia Editore, Caltanissetta–Rome

Rossi Doria B (2003) La Sicilia: da Regione del Mezzogiorno a periferia dell'Europa forte. In: Lo Piccolo F, Schilleci F (eds) A sud di Brobdingnag. L'identità dei luoghi: per uno sviluppo locale autosostenibile nella Sicilia Occidentale. FrancoAngeli, Milan, pp 11–41

Rossi Doria B (2007) La Sicilia: una regione di città. In: Rossi Doria B (ed) Sicilia terra di città. IGM, Florence, pp 11–26

Rossi Doria B (2009) La Conca d'Oro. I processi di urbanizzazione, le aree agricole, le politiche e i piani a Palermo. In: Leone M, Lo Piccolo F, Schilleci F (eds) Il paesaggio agricolo nella Conca d'Oro di Palermo. Alinea, Florence, pp 25–38

Rossi Doria B, Lo Piccolo F, Schilleci F, Vinci I (2005) Riconoscimento e rappresentazione di fenomeni territoriali inediti in Sicilia. In: Proceedings of the 9th SIU conference 'Terre d'Europa e fronti mediterranei', Palermo, 3–4 March 2005, vol 1. Zangara Editore, Bagheria, pp 263–273

Schilirò D (2012) Industria e distretti produttivi in Sicilia fra incentivi e sviluppo. StrumentiRes 4(1):1–10

Schilleci F (2005) Il contesto normativo in Sicilia. Una difficile pianificazione tra ritardi e resistenze. In: Savino M (ed) Pianificazione alla prova nel mezzogiorno. FrancoAngeli, Milan, pp 189–208

Schilleci F (2008a) La dimensione metropolitana in Sicilia: un'occasione mancata? Archivio di Studi Urbani e Regionali 91:147–161

Schilleci F (2008b) Visioni metropolitane. Uno studio comparato tra l'Area Metropolitana di Palermo e la Comunidad de Madrid. Alinea, Florence

Scott AJ (2008) Social economy of the metropolis: cognitive-cultural capitalism and the global resurgence of cities. Oxford University Press, Oxford

Soja EW (2000) Postmetropolis: critical studies of cities and regions. Blackwell, Malden, MA

Soja EW (2011) Regional urbanization and the end of the metropolis era. In: Bridge G, Watson S (eds) New companion to the city. Wiley-Blackwell, Chichester, pp 679–689

Trigilia C (2012) Non c'è Nord senza Sud. Perché la crescita dell'Italia si decide nel Mezzogiorno. Il Mulino, Bologna

van den Berg L, Drewett R, Klaassen LH (1982) Urban Europe: a study of growth and decline. Pergamon, Oxford

Forms and Processes of Settlement Pressure on Natural Systems

Filippo Schilleci, Annalisa Giampino, and Vincenzo Todaro

Abstract The process of urbanisation is considered one of the most significant anthropic alterations of the environmental framework, and the present study attempts to understand the spatial characteristics of urban growth and their impacts on environmental components in two metropolitan areas of Sicily. Such rapid and unplanned urbanisation in Sicily has increased during the last three decades causing severe pressure on various natural resources. In this situation, the level of fragmentation and isolation of natural areas risks becoming worse. The question assumes a particular relevance in those *local* territorial portions, such as some urban and periurban areas, presenting relevant conditions of *naturality* close to highly built environments. Such landscapes are usually affected by forms of human pressure that have determined intense deterioration processes, that urban and regional planning tools were not able to control and reduce. Empirical evidence of this study suggests that alternative urban patterns generate differential effects on the environmental system. Starting from this evidence, in this article we explore the reciprocity relationship between built environment and open territory, with the aim of identifying possible strategies to manage the phenomena of urban sprawl in the metropolitan context.

1 Introduction

The regionalisation processes of the urban area, in their agglomeration and explosion dynamics, mark the end of the city as a uniquely-defined morphological and functional unit. They produce vast, highly-fragmented urbanised territories, where

F. Schilleci (✉) · A. Giampino · V. Todaro
University of Palermo, Palermo, Italy
e-mail: filippo.schilleci@unipa.it

A. Giampino
e-mail: annalisa.giampino@unipa.it

V. Todaro
e-mail: vincenzo.todaro@unipa.it

© The Editor(s) (if applicable) and The Author(s), under exclusive
license to Springer Nature Switzerland AG 2021
F. Lo Piccolo et al. (eds.), *Urban Regionalisation Processes*,
UNIPA Springer Series,
https://doi.org/10.1007/978-3-030-64469-7_3

the plurality of settlement logics is manifested in a varied syntax of voids and solids, interstitial agricultural spaces and service spaces, natural areas and residential and industrial spaces. The spatial expansion of this complex mosaic of settlement situations, however, multiplies the negative effects on environmental systems that are generated by anthropisation processes.

With reference to post-metropolitan territories, the new settlement phenomena often take the forms of low-density dispersed urbanisation, which generates many problems in terms of impacts on natural and semi-natural areas. Trying to schematise what is in fact the subject of a broader debate, we can identify the following factors among the major environmental consequences of low-density urbanisation processes:

- The consumption of soil, understood not only as an irreproducible and finite resource in itself, but also in relation to the hydrogeological risks associated with the waterproofing of soils and cementing;
- The fragmentation of ecosystems (EEA 2006), generated by the subtraction of soil from agricultural activities and natural areas;
- The increase in pollution, due to the growth of private mobility.

In relation to the problems highlighted, this Chapter proposes the analysis of the main forms of settlement pressure on natural systems in relation to the reciprocal spatial relationships between the built environment and open territories. Furthermore, these reflections question the possibility of identifying alter-urbanisation strategies (Brenner 2016) and of governing settlement dispersion phenomena.

2   Soil Consumption and Urbanisation Phenomena

The consumption of the soil due to urbanisation is a phenomenon that has already been under observation for many years, but of which only the scientific environment has noticed the growing high environmental impact. From a disciplinary point of view, it is possible to identify a line of research focusing on the analysis and identification of new morphologies of urbanisation and on the interaction among different settlement patterns coexisting in a specific context. This interpretation complements scientific production on the forms of settlement pressure on environmental systems, producing change and deterioration in natural ecosystems and loss of biodiversity. One necessary reference for this line are the studies on fragmentation processes (linear, concentrated, mixed) (Romano 2002) and those on the seclusion of natural and semi-natural habitats (protected or unprotected). Such references combine with the analyses conducted on the process of replacing natural systems with artificial ecomosaics.

The state of urbanisation in Sicily is affected by the phenomenon of urban sprawling which, starting from the Seventies, has strongly contributed to shaping the regional territory both from the physical and functional points of view (Fig. 1). Such a phenomenon has mainly affected the fringe areas around the metropolitan areas where land consumed by low-density settlements amounts to 42% of the entire urbanised territory; such a percentage, when considered for coastal municipalities, is over 75% of the urbanised territory (INU 2003).

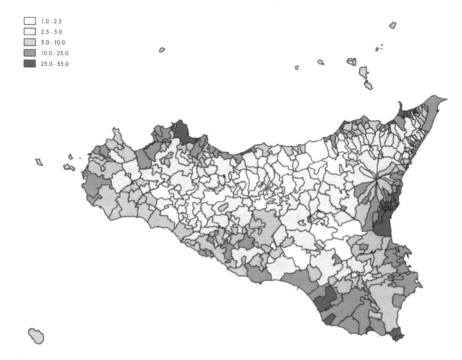

Fig. 1 Land use in Sicilian municipalities (% of Land Area) (Giampino 2018, 30)

In the case of natural systems, human pressure phenomena weigh on an already critical situation, caused by a spotted distribution of protected areas. The level of fragmentation and isolation of these areas risks becoming worse. In Sicily, the absence of environmental assessment tools and the disconnection between urban and regional planning on one side, and sector planning on the other have determined an inadequate control level off human pressure on environmental systems.

Regarding this general framework, this article clarifies and reformulates the reciprocity relationship between built environment and open territory, with the aim of identifying possible strategies to manage the phenomena of urban sprawl in the metropolitan context, a main objective of European territorial policies (EEA 2006) and of many national policies of the member states.

The consumption of soil due to urbanisation (a form of usage to be considered *irreversible* in comparison to those entirely efferent to the activity of settlement which include agriculture in its various manifestations) is a phenomenon already for many years under observations, but of which only the scientific environment has noticed the elevated environmental impact progress. On an international level, mostly in the USA, the signalling of attention to the so-called urban sprawl and the relative initiatives of studies has been given for many years (Buttenheim and Cornick 1938; Haskell and Whyte 1958; Mumford 1961; Gaffney 1964; Altshuler 1977; Mitchell 2001). In Europe, the problem of the velocity of increments of the urbanised spaces

(Hall 1998) has brought countries like Great Britain and Germany to use the well-known instruments of control of the phenomena based on rigorous limitations of the new buildings.

From a disciplinary point of view, the research on cognitive issues about the contemporary settlement dynamics have brought about the elaboration, specifically from the early Nineties, of some multiple interpretations wherein different situations have been concerned (illegal settlement, planned lotting, industrial and commercial settlement, etc.). There is a trend of studies found in this research dealing with the analysis and the identification of the new morphologies with regards to urbanised soil and its typological articulation (Galster et al. 2001; Ewing 2008).

The current study refers to this theorical mark, trying to provide a partial progress to a debating issue, still unsolved in its own presupposition or leading questions, researching the new urbanised forms in comparison with the effects that they produce on the environmental structure. From a methodological point of view, by means of a case-study, focused on a multiple scheme, an analytical-interpretative plan has been drawn up to return the local declinations of the urban phenomenon.

Catania and Palermo, two of the most important Sicilian metropolitan areas, have been selected, so as to identify the metropolitan area as a general researching issue—as by enacted law—and not intended as a territorial system which effectively gravitates round a metropolis. As per the research topics, a morphological study about the new urbanisations has been carried out, dealing with the interpretation of the geometrical configurations generated between settlement and infrastructure and the effects that they produce, in terms of fragmentation, on the agricultural, environmental and naturalistic systems. A preliminary literary recognition was necessary so as to introduce the issue of the emerging forms of urbanisation and environmental fragmentation.

The first attempts to codify the new forms of urbanisation that occurred in the early post-war period, in the USA, are a consequence of the clear manifestations of the anarchical and unlimited growth of the periurban areas. In a contest of great alteration, while the question of urban explosion had reached alarming levels, the theoretical elaborations defined *interpretative figures*, which remained idiomatic, such as megalopolis (Gottmann 1970) and urban sprawl.

In Europe the disconnection between the traditional categories of analysis and the urban phenomenology of the territorial dimension began in the Sixties. The Fifties and the Sixties characterised a period of great changes in terms of transformation of territorial and urban geographies, creating, in less than a decade, a new urban form known as the metropolitan city, whose growing dynamics record a rapidity of modification never known before in urban history. It is the age of the supremacy and proliferation of the urban element which, in terms of discipline, returns the formulation of neologisms intended to identify the new urban typologies. Some expressions such as *regional city* (De Carlo 1962), *urban diffusion* (Ardigò 1967), *linear-city* (Soria y Mata 1968), *diffuse city* (Indovina et al. 1990) or *edge city* (Garreau 1991), *urbanised country* (Becattini 2001), *ecopolis* (Magnaghi 1989), *urban bio-region*

(Magnaghi 2010) represent the attempt to catch the great modification of the territorial setting, traditionally dualistic, and the greater integration between urban areas and open space.

In Italy, research on the urbanisation processes move from a quantitative approach—based on the parameter of the soil, in terms of *measurability* and empiric *objectiveness* (Borachia et al. 1988; Astengo and Nucci 1990)—to an approach which is capable of translating the quantitative datum into a qualitative knowledge of the process (Indovina et al. 1990; Secchi 1995; Clementi et al. 1996).

In accordance with this radical change of perspective, the physical *form* becomes a transversal reference which implies all the territorial components; an explicative variable of the urban phenomenon as a result of multiple and transversal factors. In accordance with an interpretative point of view, although it is possible to consider the great effort in updating the quantitative analytical categories, although inadequate for returning the complexity and heterogeneity of the process, nevertheless, at the same time, more than the founding value of this research, we may agree with what Bianchetti (2000) says, while affirming that the issue can be considered conclusive from an interpretative point of view, only marginally entering the more relevant questions. In fact, only recently, the relationship in Italy between urbanisation processes and environmental structure has been dealt with in terms of method (Magnaghi 1989, 2010; Gambino 1997; Battisti and Romano 2007).

According to the characteristics and modalities in which these relationships are usually identified, they tend to generate pressures upon the natural component: the increasing settlement models, closely connected to infrastructural network, have determined the progressive ecological-environmental fragmentation of the natural and seminatural ecosystems.

As regards the international debate, this interpretation closely examines those forms of pressures strictly connected to the risk of biodiversity loss, which is the most influential in relation to the fragmentation phenomena (Stanners and Bourdeau 1995; Romano 2002; Battisti 2004; Hanski 2005; EEA 2006; Didham 2010; Milder and Clark 2011). More precisely, the scientific literature tends to distinguish and carries out the fragmentation process of the natural and seminatural ecosystems as a reduction of their extension and consequently as an insularisation of their environments to the point of reducing them in residual areas and changing the ecological-functional relationships among the peculiar species of a community. More than the bio-ecological component (habitat and species), the fragmentation phenomena determine relapses into the widest spatial and territorial processes and into the landscape (Jaeger and Madriñán 2011) whose matrix reference records the habitat alterations as a consequence of the reduction and disappearance of the natural vegetation.

Within the limits of the discipline, several studies have dealt with the structural impact and the eco-system functionality suggesting different solutions for the governance and planning of the territory (Jongman 1995; Collinge 1996; Romano and Tamburini 2006; Termorshuizen et al. 2007; Rudd 2012).

In Italy, the project Planning in Ecological Network (PLANECO) studied the evolutive dynamics of the ecological settings within the limits of the territory government instruments. Beginning from the ecosystemic conservative requirements and by means of the content and methods redefinition of the planning process, the project suggested an index system of environmental fragmentation which lead to individuate different forms of fragmentation according with the dynamic transformations of the anthropic (matrix) territory.

In 2003 the Agency for Environmental Protection and Technical Services (*Agenzia per la Protezione dell'Ambiente e per i servizi Tecnici, APAT*), providing a planning instruments addressing issue, methodically ordered the first considerations about the ecological networks on a local scale as an instrument to get over the fragmentary conditions of the environment. In 2011 the same institute came out with a research on the territorial fragmentation caused by linear infrastructures. This research was intended to provide good practices and ways of working just to prevent or mitigate the effect of the impacts (ISPRA-INU 2011).

At least, as regards the reciprocity relationship between the phenomena of fragmentation and the ecological connectiveness of the territory, Battisti (2004) and Battisti and Romano (2007), throughout a multidisciplinary approach which tries to reach a synthesis between languages and methods, offer a significant wide point of view about the theorical-conceptual and operative profiles which lead to the most recent experiences as regards the ecological network regulations inside the instrument planning.

The current patterns of settlement sprawl have determined an arrangement of the national territory that is heavily affected by forms of human pressure (demographic, relative to human settlement, infrastructural, productive). Those forms mainly weigh on valuable farming land, fringe areas of transition among urban agglomerations and natural and semi-natural systems, protected or not. These conditions reach the most critical level along the coastal strip, where the load due to human settlement and tourist facilities reaches its maximum, also presenting cases of illegal land occupation. In such metropolitan contexts, the phenomena of human settlement pressure have led to a significant deterioration of the traditional landscape mosaic, not only in *special* areas of natural interest, but generally in all the components characterising the landscape-environmental matrix.

3   Fragmentation Forms and Sprawl Processes on the Sicilian Territory

Sicily ranks among the 15 Italian regions that exceed 5% of regional territory consumption, with percentage values of soil consumed in 2016 which stand at 7.18%. At the same time, this is the context in which the greatest increase in land consumption was recorded on a national basis from 2015 to 2016. National Institute for Environmental Research and Protection (*Istituto Superiore per la Protezione e la Ricerca Ambientale, ISPRA*) (2017), compared to the 2014 Report, revised the estimates of the consumption of regional territory (going from values between 6.8% and 10.2% to 7.8%), while National Institute of Statistics (*Istituto Nazionale di Statistica,*

ISTAT) (2015) confirms—through the comparison of the territorial bases in 2001 and 2011—the *positive* trend of growth of the artificial surfaces of Sicily compared to the national context, with increases of 11.1% relating to the construction of new agglomerations. In essence, it is clear that the positive and constantly growing trend of the anthropisation process in progress in the Sicilian territories occurs in the face of both a decrease in the population and the recessionary trend in the construction sector.

The data on soil consumption by altitude classes (ISPRA 2017) confirm Sicily as the region with the highest percentage of soil consumed along the coastal strip (Fig. 2), with a percentage increase in soil consumption from 2015 to 2016 in the range from 0 to 1,000 m from the coastline equal to 0.45%. On 1,088 km of coastline, 662 km are urbanised (Legambiente 2016).

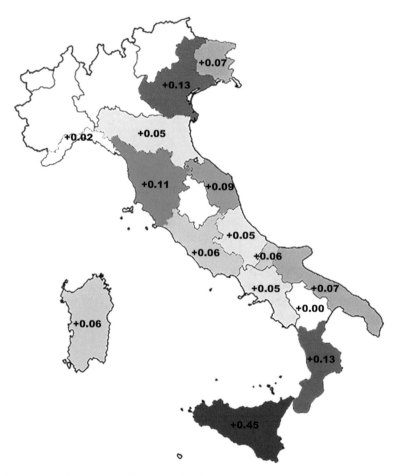

Fig. 2 Percentage increase in soil consumption from 2015 to 2016 in the coastal strip of 300 m (Image by Annalisa Giampino)

Particularly significant is the fact that 90.2% of these urbanisations were built before 1988 (Legambiente 2016), although already in 1976 the Regional Government approved the RL 78, imposing—in advance of the national provisions—a constraint of absolute no construction within the range of 150 m from the coast. Just as nothing has proved the effectiveness of the RL 37/1985, which excluded buildings illegally built there after 1976 from the building amnesty procedures. The spatial outcome of this abusive practice is a coastal strip with a density of well over 700 buildings per square kilometre and, from a penal and administrative point of view, the detection of an annual average of 400 offenses in state-owned maritime areas (Legambiente 2012). Settlement sprawl has determined, due to its form and modalities of manifestation, a significant change in the traditional Sicilian settlement, mostly characterised by compact and/or concentrated fabric, with the exception of a few specific urban contexts (Marsala network settlement system, network systems of minor municipalities on Nebrodi Mountains and Mount Etna, etc.) (INU 2003). Among the main causes of the pressures on periurban landscapes and of the consequent forms of fragmentation given by human action, it is possible to identify the following:

- Settlement phenomena (concentrated, linear, sprawled, isolated);
- Infrastructural phenomena related to mobility (simple and complex);
- Productive phenomena (areal and punctual, primary, secondary and tertiary).

3.1   Settlement Phenomena

Low construction density and intense infrastructuring of the territory, diffusely present in forms of settlement sprawl, are among the main causes of environmental fragmentation of the periurban landscape. Such pressures, in many cases, form localised and/or diffused barriers, generating a heavy interruption of the physical continuity between urban environment and extra-urban territory: on a vast scale, with the effects on the landscape (Battisti 2004) and, at local level, with the loss of physical relations between empty urban areas and open extra-urban spaces.

The sprawl urbanisation is a soil occupation typology that we may define as *pathologic*, as it implies:

- Discontinuity of urbanisation;
- Functional and social segregation;
- A consumption of territorial resources which is not justified by dynamics of demographical and occupational growth;
- High collective costs relating to need of infrastructures and service;
- Space homologation, etc. (Camagni et al. 2002).

Trying to simplify the subject of a larger debate, among the main environmental consequences of settlement sprawl, we can identify the following factors:

- Consumption of soil, not only considered as a not replicable and finite resource but also in relation to the hydro-geological risk given by the sealing of the soil and by the transformation into a concrete jungle (it is not by chance that in 2006 the "Proposal for a directive about a framework for soil protection and for modification of the Directive 2004/35/CE" was presented to the European Parliament; the proposal not only considers the risks caused by the progressive sealing of the soil, and in general all the forms of pressures on it, but it also denounces a significant increase of the soil deterioration processes caused by urban development);

- The fragmentation of ecosystems (EEA 2006) generated by the subtraction of land from agricultural activity and from natural areas;
- The increase of pollution, caused by the growth of private mobility.

Particularly in Sicily, the settlement component recorded, between 1961 and 2001, an increase of about 1.2 million houses. At the same time, within the whole built environment, the reduction from 90 to 70% of occupied houses is to be explained by the increase of the number of holiday homes, in the vicinity of the periurban zones and particularly along the coast (INU 2003; Trombino 2005). Settlement pressure conditions exist in fringe rural areas (often characterised by the presence of traditional crops of historical and cultural importance, such as citrus orchards, olive groves, vineyards), and on transitional periurban areas (particularly in the Metropolitan Areas of Palermo and Catania).

Over the last few decades, in fact, agriculture has been on the one hand characterised by a relevant loss of competitiveness, and on the other hand it has resulted in compromising large portions of agricultural land, causing the irreversible loss of its character and specificity.

The question assumes a particular relevance in those *local* territorial portions, such as some urban and periurban areas, presenting relevant conditions of *naturality* close to highly built environments. Such landscapes are usually affected by forms of human pressure that have, in recent years, determined intense deterioration processes, that ordinary planning tools were not able to control and reduce. Moreover, they are affected by a process of environmental fragmentation that has generated a progressive reduction of open surfaces and spaces and the increase of their isolation to the advantage of *uncontrolled* growth of the settlement and infrastructure system.

In the areas of special regimes (parks, protected areas, community sites, UNESCO sites), and considering them as such, all those for which—thanks to their natural and cultural values—the law foresees a specific management plan, the settling pressure is, by law, moderated and returned to more accurate qualitative forms.

In the immediate vicinity, to the contrary, marginal effects are produced on average in a more or less pronounced way and with a proportional intensity to the attractive capacity of interest of visits that the area expresses.

The increasing value of real estate is due to the increase in demand for holiday homes and hotels and the improvement of the infrastructure to facilitate the start of the *noble* resources, causing the *insular effects* in a greater way in protected areas of various kinds; this effect, even if of little importance, in the case of historical cultural heritage, assumes a substantial weight for those of a natural character.

The major problem in these cases is connected with the definition of the interaction criteria between the special planning tools which regulate the inside of the protected areas and those outside the geographical limits for the operative liberty by the local administration in the *territorial crown* of parks, reserves, architectural areas, cultural sites.

The action of limiting settlement pressure becomes therefore a prerogative of special planning and coordination on a wide scale, which has to, however, be able to pick up and measure the marginal effects to put into action the adequate measurements utilising the widespread paradigm in the broad environmental dimension.

In Sicily, settlement pressures weigh heavily on a system of protected areas, formed by 4 regional parks (Etna, Nebrodi, Madonie, Alcantara), 89 natural reserves, 235 Sites of European Importance and Special Protection Areas (SPAs) (about 23% of the regional territory). Such a system, already presenting a weakness in its dotted territorial pattern (Pinzello 2006), is in danger of being further fragmented and irremediably damaged. In such territorial contexts, the phenomena of human settlement pressure have led to a significant deterioration of the traditional landscape mosaic, not only in *special* areas of natural interest, but generally in all the components characterising the landscape-environmental matrix.

3.2   Infrastructural Phenomena Related to Mobility

Linear transportation infrastructure and productive plants have been located on the Sicilian coasts (just like on the majority of the Italian coastline). Furthermore, they have been the main place where the biggest settlement, tourist and seasonal load has been positioned, triggering problems of coastal protection and sea water pollution.

3.3   Productive Phenomena

The situation becomes particularly critical in the territories around the Industrial Development Areas localised in the vicinity of natural areas of interest.

Regarding commercial activities, they are mainly located along the main communication routes, forming linear settlements. Samples of this typology can be found along the national road 113 close to Carini, where industrial activity has been progressively substituted by shopping malls.

These sprawl modalities have reached such a level of soil consumption (including agricultural and valuable land) as to jeopardise not only the landscape and environmental value of the affected territories, but even their identity and productivity.

Considering the traditional lack of specific design action provided for sprawl territories, caused by the disciplinary difficulty of definition of such spatial category, we see that there is no correct answer either in the design strategies internal to the traditional planning tools, or in new specific policy tools.

All this is in contrast with art. 2 of the European Landscape Convention (July 2000), which applies preservation policies to natural, rural, urban and periurban areas, with the aim, besides that of protecting the landscape, of its management and planning. These conditions all contribute to generate a highly-fragmented and generally analogous landscape.

4 Comparing Metropolitan Areas: Palermo and Catania

The settlement growth patterns that characterise the phenomena of urbanisation taking place in Sicily have caused several types of anthropic pressure (demographic, settled, infrastructural, productive), which, particularly in the metropolitan areas, lie heavily on natural and semi-natural systems, whether protected or not.

Particularly compared with the latest ones, the anthropic pressure phenomena and the related forms of fragmentation insist on an existent critical territorial distribution of the *here and there* protected areas (5 regional parks, 89 natural reserves, 235 Special Protection Areas (SPAs) and Sites of Community Importance (SCIs), approximately 23% of the regional territory) whose level of fragmentation and isolation is in danger of getting worse (Schilleci 2008). Besides, in reference to the agricultural landscape, although on a provincial scale, in the last decade the Utilised Agricultural Area (UAA) in Palermo has grown from 236,764 ha (2000) to 266,362 ha (2010) and in Catania from 146,213 ha (2000) to 169,274 ha (2010), it is mainly within this context that we have recorded the most serious forms of fragmentation and insularisation of the open spaces.

These conditions reach particular levels of criticality along the coastline where most of the anthropic demand connected to the settled uses is concentrated (five of nine provincial capital cities, including three metropolitan areas headquarters are located along the coast), with obvious cases of unlawful occupation of land (Trombino 2005).

4.1 The Metropolitan Area of Palermo

In relation to the outlined framework, the Metropolitan Area of Palermo has a heritage of natural areas, archaeological sites, biotypes of great importance, inducing interpreting its territory not so much through the administrative demarcation, but as part of that broader environmental system that affects the entire region.

In this regard, the texture of areas of high environmental value is represented by twelve natural reserves, imposed by RL 98/1981, and subsequent amendments thereto, from forty-six SCIs and three SPAs and four Special Areas of Conservation (SACs). They have been identified by the implementation of "Habitats" 92/43/EC and "Birds" 79/409/EEC Directives. In addition, they fall back inside two protected marine areas: the Ustica marine area set up with *DI* of 12 November 1986 and that of the Cape Gallo-Isola delle Femmine established with *DM* of 24 July 2002.

Fig. 3 Spatial pattern of division between the coastline and the natural system Cinisi and Terrasini (Image by Vincenzo Todaro)

In addition to the presence of areas of natural interest, subject to different protection schemes, it is possible to trace an interesting system of valuable agricultural areas (environmental and landscape interest), which contribute to outlining the identify value of this area. In particular, it addresses two prevailing agricultural landscapes: the orchards and the olive groves. The former stretches along the coastline between Palermo and Trabia, while the latter stretches from the internal areas of Altofonte up to the last coastal fringes of Termini Imerese. The agricultural system of the vineyards that extend from the Western fringes of the Metropolitan Area towards Alcamo (in the Province of Trapani) must be added to these landscapes.

Related to the settled system, it is possible to identify three sub-systems that respectively fall back around three urban focuses of the metropolitan area: Palermo, Carini and Termini Imerese. The area of Palermo is mostly affected by the presence of typical residential scattered development which denies any form of relationship with the road system. The Carini context is characterised by the widespread presence of leapfrogging development, pronged and unclear type without interruption, stretching from the far northern suburbs of Palermo up to the municipalities of Partinico and Balestrate. This area has a high degree of complexity, in relation to the presence of a productive/commercial district running along the national road 113 raising the linear typical shape.

In the Termini Imerese area, unlike Carini, we can see a greater presence of pronged settlements which stand on historical paths of roads connecting the residential areas, while—as happens in the Carini area—in the area stretching between the national road 113 and coastline, the residential urbanisations assume a linear configuration, as the presence of parallel productive/commercial blocks are found, in this case, to the coast and bordered by infrastructural ways formed by the highway, the state highway and the railway line. This location results in a fringing fragmentation of wide-range environmental impact which has in fact distorted the coastal scenery of this area, producing high levels of division between the coastline and the natural system (Fig. 3) characterised by the Oriented Nature Reserve (*Riserva Naturale Orientata, RNO*) of Mount San Calogero.

In the western part of the same sub-context—stretching between the towns of Altavilla Milicia and Trabia—the low-density residential linear strip settlement (originally included between the railway line national road 113 and the motorway and which later expanded beyond the highway route) has generated increasingly rarefied forms which determines levels of complex fragmentation in comparison with both foothill agricultural surroundings and the natural system in which the SCI "Mount Cane, Pizzo Selva a Mare, Mount Trigna falls" is located (Fig. 4).

In the framework of Carini, the presence of different forms of urbanisation—associated with the productive/commercial block and the infrastructure parallel to

Fig. 4 Spatial pattern of settlement pressure on SCI "Mount Cane, Pizzo Selva a Mare, Mount Trigna falls" (Image by Vincenzo Todaro)

the coastline and within the Plain of Carini—determine a single macro-system which generates multiple types of fragmentation. In fact, you can find both forms of pressure on agricultural systems, which are nowadays residual ones, and on the highest natural environmental systems.

4.2   The Metropolitan Area of Catania

The consistency of the areas of natural interest consists of three regional parks which are in the provincial territory (Parco dell'Etna, Nebrodi Park which lies partly in the Provinces of Messina and Enna, Alcantara Park which also lies partially in the Province of Messina), seven nature reserves established by RL 98/1981, and later related changes, by thirty SCIs by two SPAs and five SCI-SPA, identified in implementation of "Habitats" 92/43/EEC and "Birds" 79/409/EEC Directives. It also falls within the Cyclops Islands natural marine reserve established by the *DM* of 09 November 2004. The Metropolitan Area also has a rich system of main agricultural areas that still contributes to defining the cultural features of this area.

It particularly deals with two prevailing farming systems, the citrus groves (along the slopes of the Etna volcano towards the Plain up to the territory of Paternò) and the vineyards (along the northern hillsides of Etna and in the area of Calatino).

From the point of view of territorial structure, the Metropolitan Area of Catania includes an urban system which is structured around the Etna volcano and is divided into four sub-systems which follow the main development axes that radiate from the centre of Catania (Dato 1991). In the coastal towns, from Catania towards Acireale, residential urbanisation takes on typical linear forms, or linear strip development directly behind the coast. In the inland areas, however, residential development has followed different routes: directly spreading on both main and secondary roads as in Paternò and Trecastagni; denying the relationship with road conditions, and assuming the typical scattered configurations as in the municipality of Misterbianco; or moreover following the paths of linking roads among the various urban centres such as Pedara, Nicolosi, Mascalucia, determining the typical branching shape.

In addition to forms of urbanisation of residential type we can find some highly specialised areas of productive/commercial type which take the form as compact development (such as the case of the Southern area of Catania) or linear trend as in the cases of Gravina or San Giovanni La Punta. The Southern productive/commercial block of Catania particularly offers an interesting observation field as it develops in a traditionally agricultural area where the industrial activity was gradually replaced by commercial structures. Besides, from the morphological point of view, the compact development shows characteristics of low density still showing residual agricultural areas currently neglected.

In terms of impact on the environmental system, the presence of the *RNO* Simeto Oasis (Fig. 5) has been detected, SCI (ITA070001) at the mouth of the Simeto River, Gornalunga Lake and the SPA (ITA070029) Biviere of Lentini, a stretch of the Simeto River and area in front of the mouth. All this leads to fragmentation on both

Fig. 5 Perimeter of RNO Simeto Oasis (Image by Vincenzo Todaro)

the agricultural areas and around the areas of natural interest that, as in the case of the mouth of the river Simeto (Fig. 6), Gornalunga Lake and the SPA Biviere of Lentini, a stretch of the Simeto River and area in front of the mouth, on which stands a further low-density residential block that stands along the coastline.

5 From a Critical Viewpoint to Plan Orientation

In relation to the conditions of particular environmental conflict detected, the current environmental heritage protection regime is inadequate. In fact, it mainly responds to a rigid model that pits protected areas against areas with the same environmental value but unprotected. This condition, in terms of territorial distribution, is found both along the coastal strip and in the internal areas.

In general terms, it is therefore increasingly necessary to integrate the current protection model with a view to promoting the establishment of an integrated environmental heritage management system. It is necessary to find the territorial specificities that the settlement pressure has not changed, just as it is necessary to reconstruct the conditions of environmental continuity (ecological network) at local and territorial level (Forman 1995; Forman and Hersperger 1997; Gambino 1997;

Fig. 6 Spatial pattern of settlement pressure on RNO Simeto Oasis (Image by Vincenzo Todaro)

Filpa and Romano 2003). This perspective makes the coordination of safeguarding policies even more necessary between the bodies responsible for the management of areas of environmental interest and the municipal administrations in the direction of activating alternative protection paths to the traditional constraint-type ones, which directly affect the policies and on territorial governance instruments (Schilleci 2005).

As is evident from the investigation on forms and processes of human pressure on Sicilian territory, settlement sprawl is generating strong negative externalities and there is a need for innovative and far-seeing solutions of territorial re-organisation.

The regional situation is rendered all more serious by: the lack of an up-to-date planning law (the Sicilian planning law was done in 1978); the poor diffusion of territorial and vast area planning (the Regional Master Plan has never been endorsed, and just one out of nine provincial capital has endorsed the Territorial Provincial Plan); the difficulty of integrating territorial policies and specialised planning tools, falling within the competence of different agencies (regional and provincial councillorship, Monuments and Fine Arts Office, port agency, free association of municipalities, etc.); the difficulty of a complete and efficient application of the environmental assessment tools regarding projects, masterplans and programs, such as Environmental Impact Assessment (EIA), Strategic Environmental Assessment (SEA), Environmental Incidence Assessment (EIncA). As recently noted by Ombuen (2009, 55),

> today the principal problems regard [...] diffusions and sprawl phenomena, with heavy unsustainability effects both at a global and at a local scale. There is a need for a public

action (and an integrated planning system) to be able to manage these new problems at the appropriate scale and with the consciousness of the complex role of the factors which determine them, starting from the technological innovation, with ITEC, with the managing sciences, logistic, and with the modifications that these innovations can produce in the expectations and in the behaviour of affected people (companies, plants and institutions).

Although it suffered very much from the anthropic pressures exerted mainly by the growth of the settled system, the heritage of the natural interest areas reported both to the context of Palermo and Catania still maintains significant levels of identity and awareness, whose value is given, however, to the individual plans and therefore is not protected nowadays adequately and organically. In both territorial contexts, in fact, neither territorial plans (Schilleci 2005), particularly at metropolitan level, nor existing protection tools related to areas of natural interest have been approved.

The close relationship between natural and anthropic systems, especially if related to urban contexts, requires a territorial planning issued by sectoral traditional approaches, to move towards integrated and broad planning and organisation able to govern the complex territorial dynamics that are related to settled forms and open spaces. In the attempt to provide directions for planning/programming tools and for the future territory government law for the control of settlement sprawl and the reduction of its impact on the environmental systems, the investigation has shown that the design solution for the sprawl territories is not to be found in banal action of building compaction, but needs to be a unitary project, made of different elements, where the empty spaces are close to compact ones, in a systemic logic supported by intermodal transportation forms. In fact, a possible control strategy for urban pressure can be implemented by establishing a system of environmental connections which connects the green areas of the city with the territorial suburban ones in such a way as to project the metropolitan contexts towards sustainable patterns of land-use planning (Schilleci 2008). This strategy underpins a dual functional value: ecological value, so that one can systematise areas of natural interest to make natural biological exchanges between them and the already existing species possible; anthropic value, so that one can enhance the system of consumption of such areas for social and recreational purposes.

Starting from these considerations, in relation to the addresses for urban planning, it is possible to identify some elements for the definition of a territory plan on metropolitan contexts (Bryant 2013), that it cannot disregard:

- Recognition, within the individual areas of urban growth, of *settlement* rules that respect and strengthen the territorial matrix (specifically it is constituted by the potential elements of ecological-environmental link and agricultural areas of advantage) resulting in structuring value;
- Pursuit of a compact city model and the concentration of its future growth along the nodes and the present infrastructures, which will attract future settlement demand, with structural effects in the overall organisation of the territory and able to reduce disorderly growth;

- Pursuit of a model of territorial development in ecological, social and productive balance, with its territory, based on the development of the specificity of the individual local nodes (Magnaghi 2010).

In this frame, within urban planning, the territorial project of sprawl territories will be based on the definition of settlement expansion models able to acknowledge, respect and strengthen the territorial matrix as a structuring element, with structural order resulting in territorial organisation to avoid messy development.

All these questions impose a reflection on the existing tools to address such territories. These tools have the limit of being disconnected from each other.

References

Altshuler A (1977) Review of the costs of sprawl, environmental and economic costs of alternative residential development patterns at the urban fringe. J Am Plan Assoc 43(2):207–209

APAT (2003) Gestione delle aree di collegamento ecologico funzionale. Indirizzi e modalità operative per l'adeguamento degli strumenti di pianificazione del territorio in funzione della costruzione di reti ecologiche a scala locale. Manuali e linee guida 26/2003, ISPRA, Rome

Ardigò A (1967) La diffusione urbana. Ave, Rome

Astengo G, Nucci C (eds) (1990) IT.URB. 80. I dati della ricerca. Quaderni di Urbanistica Informazioni 2(8)

Battisti C (2004) Frammentazione ambientale, connettività, reti ecologiche. Un contributo teorico e metodologico con particolare riferimento alla fauna selvatica. Provincia di Roma, Assessorato alle politiche ambientali, Agricoltura e Protezione civile, Rome

Battisti C, Romano B (2007) Frammentazione e connettività: dall'analisi ecologica alla pianificazione ambientale. Città Studi Edizioni, Novara

Becattini G (2001) Alle origini della campagna urbanizzata. Bollettino del Dipartimento di Urbanistica e Pianificazione del Territorio 1–2:63–69

Bianchetti C (2000) Dispersione e città contemporanea. Percorsi, linguaggi e interpretazioni. Territorio 14:161–170

Borachia V, Moretti A, Paolillo P, Tosi A (eds) (1988) Il parametro suolo: dalla misura del consumo alle politiche di utilizzo. Grafo, Brescia

Brenner N (2016) Stato, spazio, urbanizzazione. Guerini e Associati, Milan

Bryant MM (2013) Urban landscape conservation and the role of ecological greenways at local and metropolitan scales. Landsc Urban Plan 76(1–4):23–44

Buttenheim HS, Cornick PH (1938) Land reserves for American cities. J Land Public Util Econ 14:254–265

Camagni R, Gibelli M, Rigamonti P (eds) (2002) I costi collettivi della città dispersa. Alinea, Florence

Clementi A, Dematteis G, Palermo P (eds) (1996) Le forme del territorio italiano. Laterza, Bari

Collinge S (1996) Ecological consequences of habitat fragmentation: implications for landscape architecture and planning. Landscape and Urban Plann 36:59–77

Dato G (1991) Caratteri storico-morfologici degli insediamenti. In: Sanfilippo DE (ed) Catania, città metropolitana. Maimone, Catania, pp 52–61

De Carlo G (1962) La nuova dimensione della città. ILSES, Stresa

Didham RK (2010) Ecological consequences of habitat fragmentation. Wiley, Encyclopedia of Life Sciences, Hoboken

EEA (2006) The Urban Sprawl: The Ignored Challenge. Report 10/2006. Joint EEA-FOEN, Copenhagen

Ewing R (2008) Characteristics, causes, and effects of sprawl: a literature review. In: Marzluff J, Shulenberger E, Endlicher W, Alberti M, Bradley G, Ryan C, ZumBrunnen C (eds) Urban ecology, an international perspective on the interaction between humans and nature. Springer, New York, pp 519–535

Filpa A, Romano B (eds) (2003) Pianificazione e reti ecologiche. Gangemi, Rome

Forman RTT (1995) Land Mosaics. Cambridge University Press, Cambridge

Forman RTT, Hersperger AM (1997) Ecologia del paesaggio e pianificazione, una potente combinazione. Urbanistica 108:61–66

Gaffney M (1964) Containment policies for urban sprawl. In: Stauber R (ed) Approaches to the study of urbanisation. Proceedings. Governmental Research center, The University of the Kansas, pp 115–133

Galster G, Hanson R, Ratcliffe MR, Wolman H, Coleman S, Freihage J (2001) Wrestling sprawl to the round: defining and measuring an elusive concept. Housing Policy Debate 12(4):681–717

Gambino R (1997) Conservare innovare. Utet, Turin

Garreau J (1991) Edge city: life on the new frontier. Anchor Books, New York

Giampino A (2018) Consumo di suolo. In: La Greca P, Vinci I (eds) Rapporto sul Territorio Sicilia. INU Edizioni, Rome, pp 27–30

Gottmann J (1970) Megalopoli. Einaudi, Turin

Hall P (1998) Cities in civilization: culture, technology, and urban order. Weidenfeld & Nicolson, London

Hanski I (2005) Landscape fragmentation, biodiversity loss and the social response. Embo 6:388–392

Haskell D, Whyte W (1958) The city's threat to open land. Architectural Forum 108:86–90

Indovina F, Matassoni F, Savino M, Sernini M, Torres M, Vettoretto L (eds) (1990) La città diffusa. Daest-Iuav, Venice

INU (2003) Rapporto dal Territorio 2003. Sicilia, INU Edizioni, Rome

ISPRA (2017) Il consumo di suolo in Italia, edizione 2014. ISPRA, Rome

ISPRA–INU (2011) Frammentazione del territorio da infrastrutture lineari. Indirizzi e buone pratiche per la prevenzione e la mitigazione degli impatti. Manuali e linee guida 76.1/2011. ISPRA, Rome

ISTAT (2015) Rapporto BES. Il benessere equo e sostenibile in Italia, ISTAT, Rome

Jaeger AG, Madriñán LF (2011) Landscape Fragmentation in Europe. Report, 2. Joint EEA-FOEN, Copenhagen

Jongman R (1995) Nature conservation planning in Europe: developing ecological networks. Landscape Urban Plann 32(8):169–183

Legambiente (2012) Il consumo delle aree costiere italiane. La costa siciliana, da Trapani a Messina: l'aggressione del cemento e i cambiamenti del paesaggio. Legambiente, Rome

Legambiente (2016) Il consumo delle aree costiere italiane. La costa siciliana: l'aggressione del cemento e i cambiamenti del paesaggio. Legambiente, Rome

Magnaghi A (1989) Ecopolis: per una città di villaggi. Housing, 3. Clup, Milan

Magnaghi A (2010) Il progetto degli spazi aperti per la costruzione della bioregione urbana. In: Magnaghi A, Fanfani D (eds) Patto città campagna. Un progetto di bioregione urbana per la Toscana centrale, Alinea, Florence, pp 35–64

Milder J, Clark S (2011) Conservation development practices, extent, and land-use effects in the United States. Conserv Biol 25:697–707

Mitchell JG (2001) Tutti in città. National Geographic Italia 7:57–79

Mumford L (1961) The city in history: Its origins, its transformations, and its prospects. Harcourt, Brace and World Inc., New York

Ombuen S (2009) Un nuovo governo insediativo per i territori metropolizzati. Urbanistica Dossier 111:53–56

Pinzello I (2006) Le aree protette in Sicilia a 25 anni dalla L.r. 98/81. Urbanistica Informazioni 208:30–31

Romano B (2002) Evaluation of urban fragmentation in the ecosystems. In: Proceedings of the International Conference on Mountain Environment and Development (ICIMED), Chengdu, 15–19 October. VR-China, pp 11–19

Romano B, Tamburini G (2006) Urban sensibility of landscape structures: general characteristics and local details in Italy. In: Proceedings of 46th Congress of European Regional Science Association "Enlargement, Southern Europe and the Mediterranean", August 30th–September 3rd, 2006, Volos, Greece. European Regional Science Association (ERSA), Louvain-la-Neuve, pp 1–12

Rudd A (2012) Landscape Connectivity and City-Region Planning. Journal of Landscape Architecture 7(2):90

Secchi B (1995) Resoconto di una ricerca. Urbanistica 103:25–30

Schilleci F (2005) Il contesto normativo in Sicilia. Una difficile pianificazione tra ritardi e resistenze. In: Savino M (ed) Pianificazione alla prova nel mezzogiorno. FrancoAngeli, Milan, pp 189–208

Schilleci F (2008) Visioni metropolitane: uno studio comparato tra l'Area Metropolitana di Palermo e la Comunidad de Madrid. Alinea, Florence

Soria y Mata A (1968) La città lineare. Il Saggiatore, Milan

Stanners D, Bourdeau P (1995) Europe's environment. The Dobr'ıs assessment, European Environment Agency, Copenhagen

Termorshuizen J, Opdama P, van den Brink A (2007) Incorporating ecological sustainability into landscape planning. Landscape Urban Plann 79:374–384

Trombino G (2005) Le coste: urbanizzazione ed abusivismo, sviluppo sostenibile e condoni edilizi. In: Savino M (ed) Pianificazione alla prova nel mezzogiorno. FrancoAngeli, Milan, pp 279–292

An Analysis of the Residential Segregation of Foreigners in the Municipality of Palermo

Annalisa Busetta, Angelo Mazza, and Manuela Stranges

Abstract Ethnic residential segregation in Italy is emerging as a key question, which will be crucial in the definition and implementation of both urban and social policies. This chapter focuses on this phenomenon in the Italian Metropolitan Area of Palermo. We use individual data for all the population residing in the city on 31 December 2011, organised by ethnicity and neighbourhood to describe the spatial distribution and the residential segregation of foreigners in the city, applying many different segregation measures. Results show that the top ten nationalities in Palermo are not uniformly spread across the different parts of the city but tend to cluster in specific areas. The most segregated nationalities are people from China, Côte D'Ivoire, and Bangladesh, while Romanians present the lowest level of segregation.

1 Introduction

Spatial segregation of foreigners within metropolitan areas, which is a complex multidimensional phenomenon, is becoming a highly important issue in the political agenda, especially since increasing immigration flows that have involved many European countries in recent decades.

A. Busetta (✉)
University of Palermo, Palermo, Italy
e-mail: annalisa.busetta@unipa.it

A. Mazza
University of Catania, Catania, Italy
e-mail: angelo.mazza@unict.it

M. Stranges
University of Calabria, Rende, Italy
e-mail: manuela.stranges@unical.it

© The Editor(s) (if applicable) and The Author(s), under exclusive
license to Springer Nature Switzerland AG 2021
F. Lo Piccolo et al. (eds.), *Urban Regionalisation Processes*,
UNIPA Springer Series,
https://doi.org/10.1007/978-3-030-64469-7_4

For many European countries, such as Italy, this issue is a relatively new one, unlike countries with a tradition of immigration, such as the US. For Italy, the segregation of migrants is an emerging key question, which will be crucial in the definition and implementation of both urban and social policies.

Spatial segregation can be described as the residential separation of some groups within a certain society. If all members of a group are uniformly distributed across a geographical area, it can be said that this group is fully distributed. The question of segregation arises as the distance from this uniform distribution grows (Johnston et al. 1986) and spatial segregation occurs when in some areas there is an over-representation of members belonging to a group of migrants who are under-represented in others.

There are two main theoretical strands as to the causes of residential segregation: the first one suggests that the phenomenon is somehow voluntary, caused by a preference of individuals for self-segregation, while the second one suggests that it is involuntary, due to economic constraints, social exclusion, discrimination in the housing market and concentration of social housing (Darden 1986; van der Laan Bouma-Doff 2007).

Individual characteristics of foreigners (regarding, for instance, their migratory experience, employment, personal skills, education, and so on) can also be factors in explaining the level of ethnic segregation (Timms 1971; Borjas 1995; Fong and Wilkes 2003; Logan et al. 2004). Moreover, existing family ties can influence residential decisions (Zorlu and Mulder 2008; Zorlu and Latten 2009; Zorlu 2009). The residential location of immigrants is influenced by levels of acculturation and socioeconomic mobility (Grbic et al. 2010), as pointed out by the Spatial Assimilation Theory (Massey 1985; Massey and Denton 1988). This phenomenon shows two opposing spatial forces: on the one hand, there is the concentration, which produces ethnic residential segregation, and on the other hand, there is the dispersion, which produces the spatial integration of ethnic groups. Thus, the process of spatial assimilation, or ethnic residential integration, occurs as minority groups acculturate and achieve socioeconomic mobility (idem).

The literature on ethnic segregation in Western countries is steadily growing: among the most recent contributions see Nielsen and Hennerdal (2017) and Malmberg et al. (2018) about Sweden; Costa and de Valk (2018) about Belgium; Massey and Tannen (2017) about the United States; Rogne et al. (2020) about Norway. Some papers concentrate on the analysis of segregation patterns in specific cities (e.g., see Tintori et al. 2018 for an analysis of segregation patterns in seven European capitals). As regards Italian cities, recent contributions include Mazza and Punzo (2016) about the city of Catania and Busetta et al. (2016) on Palermo.

In the present work, we will focus on the phenomenon of spatial segregation of foreigners in the city of Palermo. For this we will use two data sources: (i) data drawn from the population records from the city *Anagrafe* (Registry Office) on 31 December 2011 and (ii) data drawn from the Census on 21 October 2011.

Table 1 Foreigners residing in the city of Palermo according to different data sources (absolute values)

Data source	Date	Number of foreigners
1. Census	09 October 2011	19,644
2. *Anagrafe* (pre-census correction)	31 December 2011	28,226
3. *Anagrafe* (post-census correction)	31 December 2011	17,494
4. *ISTAT* post-census correction	31 December 2011	19,804

Authors' elaboration

We have decided to integrate these two sources because they cover two different areas (units) of analysis: district (the so-called Italian *quartiere*) for the Anagrafe data and census tracts for the Census data.

For each of these two sources, we have the population by ethnicity residing in every area of the city. We will describe the residential segregation of foreigners by means of traditional techniques of analysis.

The remainder of the paper is organised as follows. In Sect. 2 we describe the two data sources, outlining the differences among them, and methods. In Sect. 3 we present and discuss the main empirical results obtained with the application of the measures proposed in the previous section, while in Sect. 4 we draw some brief conclusions.

2   Data and Methods

2.1   Data

For the purpose of this paper, we use two data sources: the administrative data from the city *Anagrafe* (Registry Office) in Palermo on 31 December 2011 (post 2011 Census correction) and the Census data. Before turning to the presentation of methods and results, it is useful to provide some details on these data sources.

As previously explained, the main difference between them is that they refer to two different areas (units) of analysis. Obviously, these sources show differences in the number of foreigners they account (Table 1).

Prior to Census correction, *Anagrafe* (source 2) accounted for 28,226 foreigners, while this value dropped to 17,494 after the census correction. Moreover, it can be noted that in the Census (source 1) there are 2,150 individuals about whom the Anagrafe after census correction (source 3) has no information at all. For completeness, we report that, although the Registry Office of Palermo has updated its archives with 2011 national census data, at the moment of their release[1] there were still some

---

[1]Population register data referred to 31 December in 2011 was received by authors in September 2015.

evident differences between these data and data available from National Institute of Statistics (*Istituto Nazionale di Statistica, ISTAT*) (source 4). In particular, *ISTAT* on 31 December 2011 counts 2,310 foreigners more than administrative registers.[2] This difference between the two sources is due to the fact that the administrative register for the city of Palermo, even if aligned with census records, is still affected by problems of underestimation and overestimation.[3]

In the following, we have decided to rely mainly on Census data (source 1) as well as providing in some cases analysis based on the *Anagrafe* data post census correction (source 2), mainly for comparison purposes and to provide additional information. Thanks to the Palermo *Anagrafe* and to *ISTAT* we have had access to anonymised data on foreigners and to Census detailed data on the main ten nationalities by census tracts, respectively.

2.2 Measures of Segregation and Localisation Quotient

There is no lack in the literature on residential segregation of measures and indexes, such as the segregation index and the exposure index (Bell 1954; Duncan and Duncan 1955; Taueber and Taueber 1965; Bell 1968; Peach 1975; Lieberson 1981; Massey 1985; Massey and Denton 1988; Wong 1993; Plewe and Bagchi-Sen 2001) which have been used more widely to study the phenomena in American cities (see, for instance, Brown and Chung 2006).

In our analysis, we will consider the following measures[4]:

- Dissimilarity Index (D): It measures the percentage of an ethnic group's population that would have to change residence for each neighbourhood or area to have the same percentage of that group as the metropolitan area overall. The index ranges from 0 (complete integration) to 1 (complete segregation). This measure is not sensitive to all the transfers of minority and majority residents between areas, but

---

[2]Data are taken from demo.istat.it.

[3]Census and administrative data provide comparable information on population size but can be affected by coverage and quality problems. In particular, data on foreigners coming from population registers could be incorrect due to both undercount for missed people (not included in the pre-census lists) and overcount caused by foreigners who left the municipalities to go abroad or move to another town without deleting their registration (i.e. included in the pre-census lists but not found at the 2011 Census). In order to check the reliability of data in terms of coverage, it has been established that population registers should be aligned every ten years with census records. Unfortunately, up to September 2015 there were still discrepancies between the two sources.

[4]Massey and Denton (1988) identify 20 different measures of segregation, classified into five conceptual dimensions: evenness, exposure, concentration, centralisation, clustering. The first three indices we calculate refer to evenness, while the last two refer to exposure. Indices of evenness and exposure are correlated (Massey and Denton 1988), but they measure different things: evenness measures do not depend on the relative sizes of the two groups being compared, while exposure measures do.

only to transfers of minority members from areas where they are overrepresented to areas where they are underrepresented (Massey and Denton 1988).

- Gini coefficient (G): It is "the mean absolute difference between minority proportions weighted across all pairs of areal units, expressed as a proportion of the maximum weighted mean difference" (Massey and Denton 1988, 285). The index ranges from 0 (minimum segregation) to 1 (maximum segregation). This measure is sensitive to all the transfers of minority and majority members between areas, not only to those between areas of over or underrepresentation (Schwartz and Winship 1980; Massey and Denton 1988).
- Entropy (H)[5]: It measures the (weighted) average deviation of each areal unit from the metropolitan area's entropy or racial and ethnic diversity, which is utmost when each group is equally represented in the metropolitan area. This index also varies between 0 (when all areas have the same composition as the entire metropolitan area) and 1 (when all areas contain one group only).
- Exposure (E): It measures the probability that a person from a minority group shares an area with a majority person, so expresses the degree of potential contact and interaction among individuals belonging to different groups, i.e. the degree to which particular groups share a common residential area.
- Isolation (I): It is the complement at 1 of Exposure index. Both measures of Exposure and Isolation depend on the relative sizes of the two groups being compared. In particular, the isolation index grows with area size and it is heavily dependent on general minority shares. They are useful as they can be more easily evaluated in terms of consequences.

All these indices show the average level of segregation for each nationality over the entire area, but none of them gives information about the intensity of segregation. For instance, two different national groups may have the same value of isolation index and this means that they are both concentrated in one single urban area. Yet this area may be in the central part of the city or on the periphery, and this may be related to marked differences in terms of the mean income level of the population, quality of life, availability of services, housing, transport, job opportunities and so on. Starting from these considerations, we have decided to calculate also a measure that can capture some spatial aspects of segregation, namely the Location Quotient (LQ hereafter), to analyse the residential segregation in the different districts of Palermo. The LQ expresses the ratio between the group proportion in the area and the group proportion in the entire city.

---

[5]The complete name is Theil Information Theory Segregation Index. It was proposed originally by Theil (Theil and Finezza 1971; Theil 1972).

3   Results

3.1   The Presence of Foreigners Within the City

Palermo is one of the fifteen Italian cities classified as a metropolis, according to its population size.[6] Its population has more than tripled since Italian Unification (1861), passing from 200,000 to just over 656,000 in 2011. The process of urbanisation developed in basically two phases: (i) a strong growth from 1861 to 1981 (with an average annual increase ranging from 9 to 20 inhabitants per 1,000 inhabitants, except for the first post-war period); (ii) a progressive decrease in the period between 1981 and 2011 (−6.9% in total).

In the first decade of the twenty-first century, Palermo showed a slightly positive (although decreasing) natural balance, and a negative migration balance, caused by a net internal migration, only partially offset by a net external migration.[7] The share of foreigners almost doubled from 2001 (less than 10,000 individuals) to 2011 (over 19,000, with an incidence of over 3% of the total population of Palermo).

Figure 1 shows the share of foreigners in each census tract over the total number of foreigners in the city, so it basically gives us a picture of how foreigners are distributed across Palermo. We can note that the highest part of the tracts (around 49%) all have a share of the total foreigners in the city lower than 0.01%, while only around 16% have a share higher than 0.07%. The remaining sections (around 35%) have a share of migrants between 0.01 and 0.07%.

Figure 2 shows the incidence of foreigners over the total population of each tract: in this case, it can be observed that 46% of tracts have a very low incidence of foreigners (<1%), around 32% of tracts have an incidence of foreigners between 1 and 7%, while 22% of the sections have a higher incidence (>7%). These tracts are mainly concentrated in the historical central area of the city (quartiere Tribunali-Castellammare and Palazzo Reale-Monte di Pietà).

In Table 2, we show the distribution of foreigners by nationalities.[8] We can see that the situation is composite: the recently settled communities (Romania, Ghana, and Côte D'Ivoire) coexist alongside others much more embedded in the territory (Mauritius, Tunisia, Philippines, and Sri Lanka). The top ten nationalities in Palermo account for more than 81% of the total number of foreigners in the city.

---

[6]In Italy art. 114 of the Italian Constitution introduces the concept of metropolitan area which takes into account demographic, urbanistic and administrative data. Metropolitan areas had been recently reformed (with the NL 56 of 7 April 2014 on "Disposizioni sulle città metropolitane, sulle province, sulle unioni e fusioni di comuni") establishing 10 Metropolitan Cities in the ordinary regions (i.e. Rome Capital, Milan, Naples, Turin, Bari, Florence, Bologna, Genoa, Venice, Reggio Calabria).

[7]In the period 2002–2010 net migration to/from abroad resulted in an increase in population between 650 and 2,200 persons per year.

[8]For ease of presentation and for the sake of brevity, we will present and discuss in detail only results for the first ten nationalities in order of size.

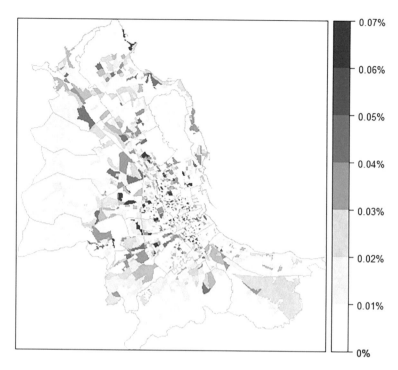

Fig. 1 Foreigners' distribution per census tract (% over total foreigners in the city) in Palermo districts in 2011 (Image by Angelo Mazza on census data, 2011)

According to census data, the largest communities are from Bangladesh (3,812), Sri Lanka (3,289), and Romania (over 1,982), followed by Ghana, Philippines, China, Morocco, Mauritius, Tunisia, and Côte D'Ivoire.

We compare these data with those stemming from the *Anagrafe*: we can note that the ranking of the first ten nationalities remains unchanged, albeit with some minor changes to the different values.

3.2 Results of the Application of Measures of Segregation

As can be seen from Fig. 3, the nationalities are not equally spread across the city districts but tend to have specific settlement patterns.

In each figure, the grey shadow in the background represents the smooth interpolation of population counts for census tracts (which becomes darker as the density grows), while each red dot represents a family with at least one foreigner. Comparing these figures, it is possible to detect differences among nationalities: in particular, China shows a strong geographical concentration in the district Oreto-Stazione and, to a smaller measure, in two contiguous districts (Tribunali-Castellammare and

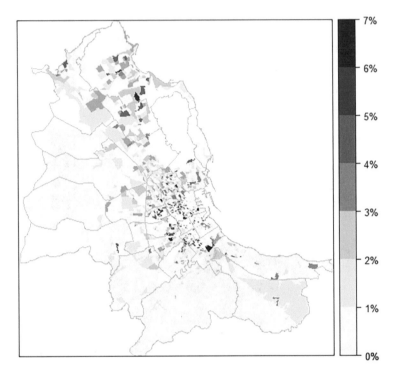

Fig. 2 Incidence of foreigners over the total population residing in each district (%) in Palermo districts in 2011 (Image by Angelo Mazza on census data, 2011)

Palazzo Reale-Monte di Pietà). The other nationalities seem more dispersed, with the Romanians as the most evenly spread across the city.

These patterns may be better understood by calculating and examining some of the segregation indexes introduced in Sect. 2 (Table 3). Although some of the indices we present in this section pertain to the same dimension as those introduced by Massey and Denton (1988), we have decided to present them all, because they may combine in many different ways and define different paths of separation of one group from another.

The empirical findings show that on average, the ethnic segregation of foreigners living in the city of Palermo is 0.7821. We then concentrate on the first ten nationalities: the most segregated nationalities are Chinese (0.9457) and Ivorian (0.9384). Chinese people live near to the central station, while Ivorians live in the historic centre, near the church of Santa Chiara.

All the other nationalities have values of dissimilarity index between 0.8215 for Sri Lankans and 0.8857 for Filipinos. The nationality which shows the lowest level of segregation is Romanian (0.6138). Romanians tend to live in all the working-class areas (where houses and costs of living are lower) spread all over the city. Women are mainly carers for the elderly, whereas men are occupied mostly in low-qualified jobs.

Table 2 First ten nationalities of foreign residents in Palermo according to the two different data sources

Country of citizenship	Census data		Anagrafe data (post census correction)	
	Absolute values	% by citizenship	Absolute values	% by citizenship
Bangladesh	3,812	19.4	3,589	20.5
Sri Lanka	3,289	16.7	3,113	17.8
Romania	1,982	10.1	1,613	9.2
Ghana	1,679	8.5	1,414	8.1
Philippines	1,337	6.8	1,259	7.2
China	1,011	5.1	913	5.2
Morocco	989	5.0	875	5.0
Mauritius	987	5.0	878	5.0
Tunisia	938	4.8	844	4.8
Côte D'Ivoire	358	1.8	307	1.8
Others	3,262	16.6	2,689	31.0
Total	19,644	100.0	17,494	100.0

Authors' elaboration on Census data (2011) and on data from the Registry Office in Palermo, 2011

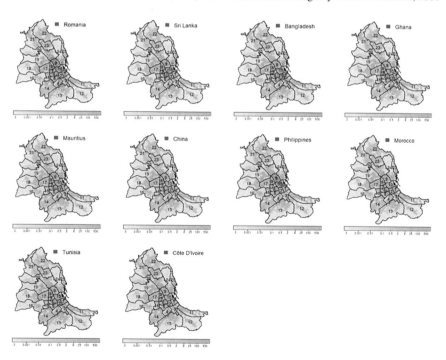

Fig. 3 Spatial distribution of households of foreign migrants (Busetta et al. 2016)

Table 3 Main segregation index for the first ten foreign nationalities

	Number of foreigners	Dissimilarity index (D)	Gini coefficient (G)	Exposure (E)	Isolation (I)
Bangladesh	3,812	0.8808	0.9614	0.0049	0.9948
Sri Lanka	3,289	0.8215	0.9178	0.0047	0.9950
Romania	1,982	0.6138	0.7848	0.0029	0.9976
Ghana	1,679	0.8281	0.9285	0.0024	0.9978
Philippines	1,337	0.8857	0.9601	0.0019	0.9977
China	1,011	0.9457	0.9806	0.0014	0.9982
Mauritius	987	0.8563	0.9394	0.0015	0.9986
Morocco	989	0.8723	0.9554	0.0014	0.9986
Tunisia	938	0.8582	0.9454	0.0014	0.9987
Côte D'Ivoire	358	0.9384	0.9746	0.0005	0.9995
Total	19,644	0.7821	0.9165	0.0058	0.9942

Authors' elaboration on Census data (2011)

The ranking of segregation by nationalities is slightly different if we observe the Gini Coefficient. According to this measure, Chinese is still the most segregated nationality (0.9806), followed by people from Côte D'Ivoire (0.9746) and Bangladesh (0.9614). The lowest value, also in this case, is recorded for Romanian (0.7848), while the other nationalities show high levels ranging from 0.91 to 0.92.

As previously mentioned, measures of evenness and exposure are correlated but measure different things. In particular, exposure indices depend in part on the relative sizes of the two groups being compared, while evenness measures do not (Iceland et al. 2014). Generally speaking, the index of isolation expresses the probability of meeting someone of one's own category in one's own neighbourhood.

In our case, it can be seen as the probability for a member of foreign nationality to reside in the same district of one of his/her own nationality. The highest value is recorded for Côte D'Ivoire (0.9995), while the lowest for Sri Lanka (0.9950).

3.3 Results of the Application of Location Quotients

As aforementioned, all the segregation indices presented in the previous section are basically aspatial, so location quotients may help to describe the spatial dimension of segregation. High values of the index indicate the strong presence of a national group in an area inhabited by other communities, while a low index would indicate the coexistence of more national groups without concentrations of a single one.

In Fig. 4 we present the distribution of the value of location quotients for each census tract by foreign nationality. Results show very different patterns by nationality:

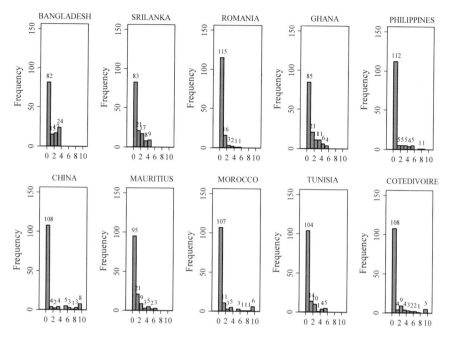

Fig. 4 Distribution of the number of districts by location quotient values for the first ten nationalities in order of size (*Note* Each histogram represents the number of census sections in which the location quotient has a value corresponding to the modalities indicated in the *X* axis (below 1, [1–2], [2–3], [3–4], [4–5], [5–6], [6–7], [7–8], [8–9], [9–10], 10 and more). *Source* Authors' elaboration on 2011 Census data)

for instance, for Côte D'Ivoire 108 tracts have a LQ lower than 1, but there are some other tracts with a higher LQ. Also, Romania has a very high share of census tracts where the LQ is lower than 1, but there are fewer tracts that have a higher value. Some other countries, such as Bangladesh, Sri Lanka, Mauritius, and Ghana show a different pattern, with a lower share of census tracts where the LQ is lower than 1 and more tracts where the value is higher.

4 Brief Concluding Remarks

Segregation is an important topic related to the presence of foreigners in urban contexts. In this chapter, we have investigated this topic using Palermo (one of the fifteen metropolitan areas in Italy) as a case study. We have presented and discussed the presence and distribution of the foreigners within the city and calculated various measures of segregation, both aspatial and spatial.

As regards the presence of foreigners, descriptive results show that in the city of Palermo there is a coexistence of recently settled communities together with other groups who have a longer migration history within the city and, hence, are more embedded in the territory.

The top ten nationalities in Palermo account for more than 81% of the total number of foreigners. In 2011, according to the Census data, the largest communities are those from Bangladesh and Sri Lanka, followed on much lower values by Romania, Ghana, the Philippines, and China. Turning to the analysis of the distribution of these foreign communities, we note that they are not uniformly spread across the different parts of the city but tend to cluster in specific areas.

The empirical findings on the ethnic segregation of the first ten nationalities living in the city of Palermo, according to the Dissimilarity Segregation Index, show that the first two nationalities in numerical terms living in Palermo present very different levels of segregation: the most segregated nationalities are people from China, Côte D'Ivoire and Bangladesh, while Romanians present the lowest level of segregation.

We obtain a slightly different ranking when applying the other measures of segregation to our data (Gini coefficient and Segregation Index): these results confirm again that Chinese is the most segregated nationality, while Romanian is the least segregated. About the Isolation Index, it clearly appears that for those nationalities which have the highest amount (i.e. Bangladeshi and Sri Lankan) the isolation index is slightly lower, while it increases as the amount of population decreases.

Our analysis is complemented by the calculation of the Location Quotients, which give certain spatial indications. Results show that, although for some nationalities (for instance, Bangladesh), there is a high level of segregation, the LQs are low in almost all the districts. This result implies that they basically live in areas where other minority groups are based.

Segregation is an important issue in migration studies because it is closely related to social and economic integration. This contribution may potentially have shed light on this topic, by describing the patterns of the phenomenon for different nationalities in an important Italian metropolis.

References

Bell W (1968) The city, the suburb and a theory of social choice. In: Greer S (ed) The new urbanization. St. Martin's, New York, pp 132–168

Bell W (1954) A probability model for the measurement of ecological segregation. Soc Forces 32:357–364

Borjas GJ (1995) Ethnicity, neighborhood and human-capital externalities. Am Econ Rev 85(3):365–390

Brown LA, Chung SY (2006) Spatial segregation, segregation indices and the geographical perspective. Popul Space Place 12:125–143

Busetta A, Mazza A, Stranges M (2016) Residential segregation of foreigners: an analysis of the Italian city of Palermo. Genus LXXI(2–3):177–198

Costa R, de Valk HAG (2018) Ethnic and socioeconomic segregation in Belgium: A multiscalar approach using individualised neighbourhoods. Eur J Popul 34:225–250

Darden JT (1986) The significance of race and class in residential segregation. J Urban Aff 8(1):49–56

Duncan B, Duncan OD (1955) A methodological analysis of segregation indexes. Am Sociol Rev 20(2):210–217

Fong E, Wilkes R (2003) Racial and ethnic residential patterns in Canada. Sociol Forum 18:577–602

Grbic D, Ishizawa H, Crothers C (2010) Ethnic residential segregation in New Zealand, 1991–2006. Soc Sci Res 39(1):25–38

Iceland J, Weinberg D, Hughes L (2014) The residential segregation of detailed Hispanic and Asian groups in the United States: 1980–2010. Demographic Res 31(20):593–624

Johnston RJ, Gregory D, Smith DM (1986) The dictionary of human geography. Basil Blackwell, Oxford

Lieberson S (1981) An asymmetrical approach to segregation. In: Peach C, Robinson V, Smith S (eds) Ethnic segregation in cities. Croom Helm, London, pp 61–82

Logan JR, Stults BJ, Farley R (2004) Segregation of minorities in the metropolis: two decades of change. Demography 41:1–22

Malmberg B, Andersson EK, Nielsen MM, Haandrikman K (2018) Residential segregation of European and non-European migrants in Sweden: 1990–2012. Eur J Popul 34:169–193

Massey DS (1985) Ethnic residential segregation: a theoretical synthesis and empirical review. Sociol Soc Res 69:315–350

Massey DS, Denton N (1988) The dimensions of residential segregation. Soc Forces 67(2):281–315

Massey DS, Tannen J (2017) Suburbanization and segregation in the United States: 1970–2010. Ethnic Racial Stud 41(9):1594–1611

Mazza A, Punzo A (2016) Spatial attraction in migrants' settlement patterns in the city of Catania. Demographic Res 35:117–138

Nielsen MM, Hennerdal P (2017) Changes in the residential segregation of immigrants in Sweden from 1990 to 2012: using a multi-scalar segregation measure that accounts for the modifiable areal unit problem. Appl Geogr 87:73–84

Peach C (1975) Urban social segregation. Longman Publishing Group, London

Plewe B, Bagchi-Sen S (2001) The use of weighted ternary histograms for the visualization of segregation. Prof Geogr 53(3):347–360

Rogne AF, Andersson EK, Malmberg B, Lyngstad TH (2020) Neighbourhood Concentration and Representation of Non-European Migrants: New Results from Norway. Eur J Popul 36:71–83

Schwartz J, Winship C (1980) The Welfare Approach to Measuring Inequality. Sociol Methodol 9:1–36

Taueber KE, Taueber AF (1965) Negroes in cities. Adline, Chicago

Theil H (1972) Statistical decomposition analysis with applications in the social and administrative sciences. North Holland, Amsterdam

Theil H, Finezza AJ (1971) A note on the measurement of racial integration of schools by means of informational concepts. J Math Sociol 1(2):187–194

Timms DGW (1971) The urban mosaic. Cambridge University Press, Cambridge

Tintori G, Alessandrini A, Natale F (2018) Diversity, residential segregation, concentration of migrants: a comparison across EU cities. Publications Office of the European Union, Luxembourg

van der Laan Bouma-Doff W (2007) Involuntary isolation: ethnic preferences and residential segregation. J Urban Aff 29(3):289–309

Wong DWS (1993) Spatial indices of segregation. Urban Stud 30(3):559–572

Zorlu A (2009) Ethnic differences in spatial mobility: the impact of family ties. Popul Space Place 15(4):323–342

Zorlu A, Latten J (2009) Ethnic sorting in the Netherlands. Urban Stud 46(9):1899–1923

Zorlu A, Mulder CH (2008) Initial and subsequent location choices of immigrants to the Netherlands. Reg Stud 42(2):245–264

Informal Practices

Housing Issue and Right to Housing in Palermo

Annalisa Giampino, Francesco Lo Piccolo, and Vincenzo Todaro

Abstract This article explores the dichotomy between interests and values in the field of the right to housing for homeless people, focusing on the informal practices of reappropriation of spaces (properties owned by the Municipality of Palermo) in the case study of Palermo. The experience we are dealing with here must be addressed in a wider context of structural and latent conflicts concerning the city of Palermo, which are strictly related to the emergency housing issue. Indeed, Palermo's housing issue is a quite complex and articulated theme, where the ineffectiveness of public policies and the increasing poverty conditions, aggravated by the economic crisis, seem to find in the various forms of the illegal occupation, as for abandoned or unused buildings, an extreme mode to *democratically* get a denied right.

1 Introduction

The renewed deterioration of the housing issue in Italy is now a generally recognised and widely documented matter (Roma 1994; Perin Cavallo 2004) also in relation to the deficit of social and housing policies that can address the matter organically. From the early 1990s, in fact, access to housing has been considered a secondary issue, perceived minimally as a quantitative matter, as it only concerns a small segment of society (Sbetti 2008),[1] and more specifically addressed as a qualitative issue in

---

[1] The demand for social housing came from weaker social categories such as the unemployed, single-parent families, young families, the immigrants and elderly.

A. Giampino (✉) · F. Lo Piccolo · V. Todaro
University of Palermo, Palermo, Italy
e-mail: annalisa.giampino@unipa.it

F. Lo Piccolo
e-mail: francesco.lopiccolo@unipa.it

V. Todaro
e-mail: vincenzo.todaro@unipa.it

© The Editor(s) (if applicable) and The Author(s), under exclusive
license to Springer Nature Switzerland AG 2021
F. Lo Piccolo et al. (eds.), *Urban Regionalisation Processes*,
UNIPA Springer Series,
https://doi.org/10.1007/978-3-030-64469-7_5

101

relation to the size of the house, considered in terms of improving the physical characteristics and state of maintenance of public residential dwellings, and in terms of improving the quality of urban space and collective services (Ombuen et al. 2000). Thus, the housing problem became a part of the wider process of urban redevelopment which during that period involved the most deteriorated parts of cities, especially outskirts and historical old town areas. A process that had in part already been started by National Law (NL) 457/1978, that designed recovery plans as implementations of the Master Plan *(Piano Regolatore Generale, PRG)*.

On the basis of these premises, in recent years the *residual* quantitative dimension and the *new* qualitative dimension of the housing issue have placed attention on urban dynamics undertaken by various municipal councils (Sbetti 2008) that have almost all prepared strategies to *limit the housing problem* through the graduation of evictions, rental subsidies and, last but not least, using the quota of public residential housing for emergency cases. The effects of the 2008 economic crisis and the consequential increase in types of urban poverty, including the serious form of homelessness, coincided with a progressive reduction in allocated state resources for the housing sector (Rigon 2008; Sampaolo 2008), proving the inadequacy of public policies pursued up to that point by the municipal councils, due to a lack of a renewed national and regional policy on the matter. In relation to the issues found, Palermo is no exception, characterised by further levels of criticality dictated by the marginal conditions that distinguish Southern Italy contexts.

The housing issue in Palermo is a complex, varied matter both in terms of extent of the demand, and in relation to the effectiveness of the public policies adopted, to the point of realistically taking on the nature of a real *emergency* (Lo Piccolo et al. 2014, 2017). However, to understand the level of emergency that has been reached, we must take a look at some of the context conditions linked with the progressive impoverishment of the urban population due to the crisis, and the structural and chronic limits of municipal policies that have been enacted in this sector.

2   Public Policies Vs. Squatting Practices for the Right to Housing in Palermo

The data calculated by the National Institute of Statistics (*Istituto Nazionale di Statistica, ISTAT*) (2017) show a relative poverty rate in Sicily in 2017 of around 22.8%, with 1,320,580 households (more than half of Sicilian households) living in conditions of deprivation.

This is the worst figure in any Italian region and, when considered together with the employment reduction of 73,568 and a 10.2% rise in the number of jobseekers (Regione Siciliana 2015), it paints a picture of the emergency situation which Sicilian households are going through. A long, complex scenario resulting from the effects of the economic crisis overlap with underlying fragilities of the Sicilian production system.

Added to this situation is a dramatic new housing emergency related to the increase in forms of urban poverty, mainly concentrated in the three large cities of Palermo, Catania and Messina.

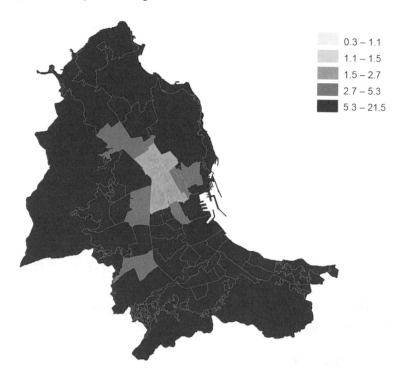

	0.3 – 1.1
	1.1 – 1.5
	1.5 – 2.7
	2.7 – 5.3
	5.3 – 21.5

Fig. 1 Incidence of families with potential economic hardship in the districts of Palermo, year 2011 (ISTAT, 8mila Census) (http://ottomilacensus.istat.it/sottotema/082/082053/15/)

Within this reference framework, the housing problem in Palermo has become significant and is affected on the one hand by long-winded processes, and on the other by recent recession dynamics that have damaged several sectors of the local economy. According to ISTAT (2011) data, the family material and social vulnerability index is 102, compared to a national value of 99.3, and with a spatial distribution of families who suffer from potential economic problems spread throughout the entire municipal area (Fig. 1). Between 2009 and 2013, 46,000 jobs were lost in the city of Palermo (Camera di Commercio and Istituto Tagliacarne 2013). Moreover, based on a recent sample survey (ISTAT-Caritas 2014) there are 3,997 homeless people[2] living in Sicily, of whom 2,887 in Palermo, confirming its third-place ranking among Italian cities in number of homeless. These are signs of economic and material problems that are heavily damaging material equality in living conditions, and the exercising of fundamental rights, in an ever-increasing bracket of the Palermo population.

The data show a status of increasingly structural material and housing deprivation, which not only affects the traditionally weaker social classes, but also groups of

---

[2]This category includes people living in extreme poverty, who over the months of November-December 2014 received food or night shelter services in 158 Italian municipalities where the survey was carried out.

Table 1 Trend data of eviction orders for residential properties in Palermo

Year	Execution orders issued				Execution requests	Evictions executed
	Evictions due to the owner's needs	Eviction upon contract termination	Evictions due to rent arrears/other	Total		
2005	2	286	1,425	1,713	1,462	532
2006	0	413	1,317	1,730	1,272	484
2007	0	311	1,133	1,444	1,467	542
2008	26	345	1,310	1,681	1,572	551
2009	5	284	1,306	1,595	2,166	719
2010	0	271	1,306	1,577	1,654	601
2011	0	159	1,222	1,381	1,519	548
2012	36	168	1,558	1,762	1,690	645
2013	1,371	130	220	1,721	1,570	639
2014	1,398	110	102	1,610	1,657	634
2015	1,324	89	107	1,520	1,324	632
2016	1,067	77	100	1,244	1,290	469

Ministero dell'Interno (2017, 71)

inhabitants for whom the public policies, that have been *parametrised* on quantitative criteria that no longer reflect the new types of poverty, cannot provide answers. It is no coincidence that 1,500 eviction orders are issued each year in the city of Palermo, with an average of 600 being executed (Table 1).

Table 1, however, shows that since 2012 there has been a slowing down of arrears situations, in favour of an increase in evictions due to the owner's needs. Although apparently this fact should indicate an overall improvement in family economic situations, this almost specular inversion is probably due to the entry into force of article 6, paragraph 5 of Law Decree (*Decreto Legge, DL*) 102/2013 (also known as the *anti-eviction decree*) that establishes and introduces blameless arrearage into Italian legal system. This is an institution created, in the wake of the crisis, with the aim of helping families affected by a clear reduction in income capacity due to the economic crisis. In fact, in the subsequent Ministry of Infrastructures and Transport (*Ministero delle Infrastrutture e dei Trasporti*) Decree of 14 May 2014, it is specified that the non-payment of rental fees, where the conditions for acknowledging blameless arrearage exist,[3] is no longer an offence and, in reference to eviction

---

[3] By blameless arrearage, the decree intends "the impossibility to pay rental fees due to the loss or consistent reduction of the family income loss of job by dismissal; company or trade union agreements with a sizeable reduction in working hours; ordinary or extraordinary lay-off that considerably limits income capacity; non-renewal of temporary contracts or of atypical work; closure of freelance work or registered companies, due to force majeure or the loss of goodwill; serious illness, accident or death of a member of the family that has brought about either a considerable reduction in total family income or the need to use a large part of income for significant medical or care costs are the main identified causes of blameless arrearage" (art. 2 of Decree 14 May 2014).

procedures, it introduces the possibility of remaining in the house by resorting to state funds managed by municipal councils. In other terms, the decree recognises family vulnerability and their progressive impoverishment, confirming the need for a joint social policy, also to the detriment of the private property market.

However, due to recent national provisions, eviction trend data for Palermo show the difficulty in applying the protective mechanism to defend social and housing problems, thus highlighting how the state has lost its capacity to mediate and control between economic and social matters (Touraine 2017). Faced with the emergency situation described, the gap between individuals facing such problems and the responses that the system provides appears to be even more dramatic. Also, the restructured welfare system has de facto transferred responsibility in this area to local councils without increasing the resources allocated to local bodies (Giampino 2012).

Therefore, faced with a housing policy that is constantly absent from the national agenda, which new municipal welfare tools has the City Council of Palermo been able to activate? What are the actions and policies enacted to address the housing emergency?

2.1 Policy Responses to the Housing Emergency

In reference to housing policies, the City Council of Palermo has operated through two intervention macro-categories: on the one hand providing economic funding; on the other allocating housing and residences and temporary shelter. Economic measures provide for three types of funding, which are alternative and not cumulative:

- The additional funding for rentals ex article 11, Law 431/1998 implemented in Chapter II of the Housing Interventions Regulations (*Regolamento Interventi Abitativi*) of the City of Palermo;
- Funding allocated to tenants with blameless arrearage ex Legislative Decree (*Decreto legislativo, Dlgs*) 102 of 31 August 2013;
- The housing poverty funding pursuant to Chapter I of the *Regolamento Interventi Abitativi* of the City of Palermo.

When looking more specifically at each type of funding, the small size of the sums available to the administration is all too clear, also due to the reduction in state and regional contributions in this welfare sector. The amount of the contribution to rent, for example, has seen no transfer of resources from the Regional Administration since 2014, to the point that since 2016 the City Council has no longer issued the relative call for applications.

Between 2009 and 2012 the funds were reduced from over € 6.5 million to under € 250,000 (specifically, from € 6,547,561.95 to € 247,409.48) (Table 2). In the best-case scenario, these figures can cover an average contribution of a mere € 400.00 per year per applicant. As the amount of the economic benefit decreased, the applications have also decreased, passing from an average of about 10,000 applications per year to 1,698 applications in the last call in 2014.

Table 2 Trend data of the contribution to rent ex article 11 Law 431/1998 of the City of Palermo

	Earmarked sums	Applications submitted	Applications granted	Applications excluded
Call 2009	€ 6,547,561.95	11,814	10,776	1,038
Call 2010	€ 5,049,613.94	13,783	12,956	827
Call 2011	€ 4,305,866.75	13,141	12,289	852
Call 2012	€ 247,409.48	10,096	6,844	3,252
Call 2013	€ 0	6,952	N/A	N/A
Call 2014	€ 0	1,986	N/A	N/A

City Council of Palermo, Bilancio sociale (2015)
https://www.comune.palermo.it/bilancio_sociale.php?anno=2015&id=24&lev=2&cap=225

This reduction is a clear sign of people's lack of trust in the measure promoted by the public body during an emergency situation. A body that has had such economic difficulties that it had to remove (from 2010 to 2016) the contribution for housing difficulties due to lack of financial coverage. This was a measure destined for extreme cases of housing deprivation. As shown, these are partial measures, and clearly insufficient for repairing a constantly growing phenomenon. And it is no coincidence that in the "Housing Emergency Ranking", provided for by Chapter V of the *Regolamento Interventi Abitativi* of the City of Palermo in relation to families affected by serious hosing difficulties (i.e. homeless or with improper housing), there are now 1,200 more families registered compared to the 858 registered in 2012. Unfortunately, this rapidly escalating situation has been tackled with inertia, with only 222 houses having been assigned in fourteen years, against 9,865 applications received with the last "Notice of open competition 2003/2004 for the allocation, in the form of a lease, of public housing units". If on the one hand the allocation of housing units is extremely slow, on the other hand the funds available to the public institutions for rent subsidies are clearly insufficient.

Lastly, faced with the creation of new housing, and in spite of available funding, the endemic slowness of the city council to complete construction of 104 houses provided for in the "Borgo Nuovo Construction Plan" (*Piano Costruttivo*) and of 32 houses created from the renovation of the former factory in Vicolo Benfante has become evident. Also, in recognising the small number of new constructions and not having any free areas on which to construct new buildings, the city council identified the renovation of existing buildings, mostly in the historic old city centre, as a possible solution to the public housing construction problem. The social urban building plan provides for the renovation of more than 1,000 buildings in the historic city centre that are at risk of collapse (with the possibility of expropriation for the owners who cannot intervene themselves) and the conversion of industrial buildings that have not been used for at least three years. Interventions that will be carried out by building cooperatives, using available regional funds to aid the sale of housing at controlled prices or offering properties at rental costs that often do not meet the economic capacity of possible tenants. It is therefore a policy that is aimed at increasing owned housing rather than implementing a social rental market for the weaker and more vulnerable segments.

In this situation, urban space has become an *object* of contention and claim by groups of inhabitants living in severe housing deprivation, who are organised at various levels, and claim—through illegal (although not illegitimate) forms of occupation of public or social private property—the right to housing as primary expression of the broader "right to the city" (Giampino and Lo Piccolo 2016).

In the capital of Sicily, roughly 600 households have occupied historical buildings, convents, schools and non-residential public buildings, also owing to the help of the said movements, and have adapted them to the new residential use, also through micro-projects of renovation which fill the lack of formal housing policies and intend to be a radical alternative to the current model.

For these informal practices, the City Council responded using a repressive policy and a *zero-tolerance* approach, then in April 2018, it proposed and obtained the inclusion of a debatable art. 72 of Finance Act (*Legge Finanziaria*) 2018 from the Sicilian Regional Assembly. An amendment that foresees the possibility of regularising the position of those who have occupied public housing prior to 31 December 2001, excluding those who have occupied non-residential buildings due to need, via housing protest committees. A provision that discriminates, not on objective criteria of need and urgency, but based on the type of public asset occupied. The partiality and ineffectiveness of this provision lies in the ambiguous nature of the squatting phenomenon occurring in places characterised by dramatic marginalisation, social and physical decay, and forms of organised crime such as the cities of Southern Italy.

2.2   Squatting Practices to Right to Housing

Squatting for living purposes, or at least the occupation acts supported by the main movements promoting the right to housing in Palermo—*Comitato Lotta per la casa 12 Luglio* and *PrendoCasa*—represent an extreme way of *democratically* obtaining a denied right (Lo Piccolo et al. 2017).

From the survey carried out on public residential buildings in Palermo, it was found that there are 19,208 properties owned by the Autonomous Council Housing Institute (*Istituto Autonomo Case Popolari, IACP*) and 4,827 properties owned by the City Council.

A small heritage compared to the public constructions built in Palermo starting from the early twentieth century and repeatedly threatened by the development and securitisation policies started up by the City Council and by the *IACP*. All this in light of a *housing demand* in a phase when the central government has delegated the market to respond to housing needs (Fregolent and Torri 2018). In Palermo, the public real estate portfolio includes 24,035 properties, of which almost a quarter is occupied illegally (5,238 properties). Specifically, we have detected that the properties illegally occupied owned by the *IACP* are 2,658, while those owned by the Municipality of Palermo are 2,580.

This is due to the fact that illegal occupation in marginal contexts, such as that in Palermo, has become contradictory and ambiguous in nature where forms of asserting a right are mixed with illegal and unlawful forms of public property occupation. In many cases, these are properties taken from possible assignees who are correctly registered in the classification and, according to estimations by the groups *Sindacato Unitario Nazionale Inquilini e Assegnatari (SUNIA)* and *Sindacato Italiano Casa e Territorio (SICET)*, over 1,000 houses in Palermo have been occupied by illegal occupants with no rights. The illegal occupation phenomenon should therefore be investigated for the ambiguity between a real, illegal market that manages occupation in return for payment, and informal practices of claiming the properties promoted by housing protest committees. This contradictory situation risks cancelling out all the real insurgent forms by those with rights, creating rhetoric and *clichés*. These interpretations tend to trace all occupation phenomena to a criminal *mafia* management and, likewise, to assimilate the squatting of the illegal occupant to that of someone colluding with this type of deviant life.

Among the buildings owned by the municipal administration, which are still occupied, is the former National Body for Italian Pensioners (*Opera Nazionale per i Pensionati d'Italia, ONPI*) in the neighbourhood of Partanna (Fig. 2). It is a retirement home built on the land donated at the end of the 50s by Baron Filippo Santocanale to the *OPCER (Opera Pia Cardinale Ernesto Ruffini)*, and for many years it has been an excellence of the territory, in terms of services offered, dimensions and characteristics.

Fig. 2 The complex of the *ONPI* in the neighbourhood of Partanna (Palermo) (Photos by Annalisa Giampino)

Covering a surface of 10,000 m², the retirement home consists of a complex of 25,000 m² split in many building having various functions: two symmetric three-storey buildings hosting the bedrooms for elderly people and common spaces; a chapel and clergy house connected to the two symmetric buildings (at present, allocated for free to the local Parish); a building used as a 200-seater theatre and a two-storey building symmetric to the theatre, where on the ground floor are located the decentralised offices of Partanna-Mondello Borough Council.

Thanks to an intervention of self-renovation supported by the members of *Aiace* Association, the retirement home was transformed into dwellings by the tenants.[4] In 2010 the complex was vacated, and the following year forty-six families (150 people in total) coming from different districts of the city transformed the spaces to adapt them to their residential needs, bearing the costs of it.

In November 2012, the small church was partially renovated; it is the heart of the complex, as well as the point for meeting and social gathering of both squatters and residents in the district. Although being aware of the ambiguity of the said experience where *legitimate occupiers* and *illegitimate squatters* coexist, the story of the *ONPI* structure, in its inception, witnesses the potential associated with self-renovation.

Nonetheless, the coexistence of different types of occupiers generates a twofold conflict: externally, between residents and occupiers, and inside the complex itself between authorised legitimate occupiers and squatters. It is not only a mere formal conflict, but it also translates into the quality of the renovation interventions: the ones having the right to do so, usually perform unrefined interventions using low-quality materials, whereas squatters perform more comprehensive interventions with higher-quality materials.

The same dynamics can be found inside the complex, in a *geography of differences*, where the contrast between the two symmetric buildings is reflected in the homogeneous settlement of the two different groups of occupiers, in the different quality of renovation and in the fortified look of the buildings occupied by squatters, unlike the ones legally occupied.

Similar experiences can be found in both the historical centre of Palermo and the outskirts of the city.

Another significant case dates back to 2014, when about thirty families occupied the former Salvemini School in Borgo Nuovo (owned by the City Council) and created apartments in the former classrooms. Another similar occupation followed this one in the same year, when 50 families occupied and self-reclaimed the former Crispi School in the *CEP* district (Coordination Committee for Public Housing, *Comitato di Coordinamento dell'Edilizia Popolare*) (Figs. 3 and 4).

In all these cases of self-reclamation of abandoned public buildings, promoted by housing protest committees, the illegal occupants were found to be legitimate

---

[4]In 1999 the bishop Salvatore Pappalardo sold the retirement home to the Municipality of Palermo. Subsequently, the retirement home was managed by many different bodies—with many co-operatives alternating rapidly—thus creating a condition of distress for the guests. On June 4th, 2010, the Municipality of Palermo, led by the Cammarata administration, after a number of attempts issued an order to vacate and transfer the elderly people living in the retirement home to other facilities, as the retirement home was judged non-conforming.

Fig. 3 Practices of adaptive use in the former classrooms of Crispi School in the *CEP* district, case 1 (Photo by Annalisa Giampino)

recipients of a Residential Public Building (*Edilizia Residenziale Pubblica, ERP*); over time, however, the lack of a public subject and any political attention has brought about—as found in these six year of monitoring occurrences—the occupation by families who are not registered on the Housing Emergency Ranking, causing conflict as shown in the former *ONPI* case-study.

This is a tangible and extreme response to the absence of a public subject and its aptitude in preserving a state of widespread illegality linked to the topic of housing rights.

The *ZEN* 2[5] (Fig. 5) experience, although owned by the *IACP*, seems to be useful

---

[5]The word *ZEN* is an acronym for *Zona Espansione Nord* (Northern Development Area), which is the name of the neighbourhood.

Fig. 4 Practices of adaptive use in the former classrooms of Crispi School in the *CEP* district, case 2 (Photo by Annalisa Giampino)

for describing the *value* of some informal practices that act bottom-up and that see citizens working to regain their own citizens' rights (Lo Piccolo et al. 2014). In particular, the phenomenon of illegal occupation appeared before the housing was completed, with the initial contribution from the institutions of silently accepting the phenomenon.

To date, according to *IACP* data, only 435 of a total of 2,894 houses are found to be legally rented, while 429 are stated as having their use status as *unknown* (therefore it is likely that they are illegally occupied). This means that 2,030 houses are officially classed by the *IACP* under the heading *illegally occupied*.

On the other hand, the fact that an official application for legalisation has been sent to the *IACP* by the occupants of 1,355 houses that are illegally occupied, is even more relevant.

Fig. 5 Aerial photo of the *ZEN* 2 neighbourhood (Bonafede and Lo Piccolo 2010, 359)

From a physical and spatial point of view, with similar dynamics to the ones found at the former *ONPI*, although here on a district scale, these data translate to a geography of marginalisation that ideally separates and divides the northern area, where the legally allocated *insulae* are located, from the southern area, where the illegal occupation and impossible to legalise situations seem to be concentrated.

At the same time, contrasting cases of the creation of communities among illegal occupants and legal assignees have been recorded, as in the case of the establishment of the non-profit organisation "Insula 3 Evolution".

One association comprising 122 families, and with its base in an illegally occupied warehouse in insula 3E, that has occupants and assignees working for the maintenance, cleaning and observance of the pedestrian areas of the *agorà* in insula 3E completed in 2012 by the *IACP*.

However, even in the controversial situation created, these insurgent acts are an opportunity to reflect on the necessary changes in housing issue action models, in the attempt to renew the possibility of a public housing policy in the changing welfare state system. In spite of this, the above-mentioned article 72, of the 2018 Finance Act, proposed by the City Council of Palermo, does not voice the petitions for renewal subtended in these acts, but are rather an undifferentiated *amnesty* that penalises those who are in the most extreme conditions of deprivation. In fact, as found by the *SICET* and *SUNIA* investigations, while the residential housing units are often occupied by those without any rights, the non-residential buildings remain the solutions for those with rights to *ERP* and who are correctly registered on the Housing Emergency Ranking. At the same time, the exclusion of the occupants of non-residential public buildings from the amnesty does not take into account the potential re-use of this type of building, which is the newest feature proposed by these actions. Currently, we need to understand the control and verification measures for the requisites that will be implemented by the city council in order to proceed with their application. Meanwhile, the possibility of adopting regulations to institutionalise self-recovery, and the bureaucracy connected with that, as a measure to aid housing problems is still a long way off.

Moreover, instead of starting a shared path to solve the problem with the homeless, the City Council has avoided the problem, tolerating illegal occupation (and thus avoiding having to issue eviction injunctions) and at the same time, in December 2015, approving an Alienation Plan for about 2,000 public residential properties and about fifty non-residential buildings.

3 Critical Issues in the Negotiation Strategies of Housing Programmes in Palermo

One last matter regarding housing policies are the negotiation practices started up by the city council in the *ERP* districts. Proposed interventions are based on bargaining models and logic and, in fact, are disconnected from the local economic and social reality, and proving to be inadequate for addressing the inhabitants' priority needs.

Starting from 1990, year in which the first season of *extraordinary* interventions due to World Football Championships funding were started, to date, a number of interventions have been started—some of which have been completed—as part of the planning tools for the redevelopment of public residential building districts that theoretically should have changed the Council's action model towards deliberative and democratic forms of construction for the city. This means about thirty years where the housing issue is not addressed in terms of availability of housing solutions, but is placed in a qualitative dimension as the result of a broader deterioration of the traditional districts produced by the Plans for Popular and Low Cost Construction (*Piani per l'Edilizia Economica e Popolare, PEEP*).[6] As Tosi (1994, 11) stated, it is "unacceptable to identify between housing problem and 'city construction'".

---

[6]The changeover from a quantitative-object logic to a qualitative-relational logic has found a practical reason in the fulfillment of the primary housing need, at the start of the 1980s, which saw about 70% of houses being owned. At the same time, these data were instrumentalised to justify a progressive reduction in public interventions in the housing sector, in favour of interventions, both tangible and intangible, on the various dimensions that are inherent in the notion of social exclusion.

Cremaschi et al. (2007) also warn of the evasive exchange committed by complex planning experiences of redevelopment and resolution of housing problems.

> The idea of returning the social housing problem to being an urban problem and handling it via redevelopment policies is an expression of a specific ideological and organisational framework of policies: a framework in which there is a high separation between social policies and urban policies and where the latter are much closer to urban-planning policies. (Cremaschi et al. 2007, 20)

With the season of complex programmes, public space in the *ERP* districts of Palermo will be contested by informal practices started by occupants and residents (with the support of local associations and movements) and official urban policies started up by the City Council. In this sense, public space is an element for comparison between mainstream urban policy based on neo-liberal and self-referential logic and the practice of bottom-up repossession and resignification started by the inhabitants. Two parallel but totally different paths that, as we will see, have almost never—and only in rare cases—met and that show the history of a city that has always been contested between reformist powers and counter-reformist movements.

3.1 Conflict and Negotiation Over Public Space in the ERP Districts

The start of negotiations on the public residential building districts was signalled by the Regional Presidential Decree (*Decreto del Presidente della Regione, DR*) of 14 December 2006, when the Integrated Intervention Programme (*Programma Integrato di Intervento, PII*) ex article 16 of National Law (NL) 179/1992 for the San Filippo Neri district (formerly *ZEN* 2) and the two Urban Recovery Programmes (*Programma di Recupero Urbano, PRU*) ex art. 11 NL 493/1993 for the Borgo Nuovo and Sperone districts were approved. These were for about 90 interventions defined as redevelopment, with a total funding of 75.7 million Euros, aimed at creating new services, infrastructures and public spaces for three *ERP* districts in Palermo, where social problems and a lack of services and public spaces become endemic parts of public building interventions in Southern Italy. In spite of the announced objective, whose primary task was to overcome the physical, urban and social deterioration of the public residential building site through a range of planning goals, and in spite of the initial declaration of a main role for public intervention, the complex plans basically favoured some private interventions on the edge of the districts as the enacting of their slow, unfinished implementation, without addressing the conditions of marginalisation, deterioration and lack of liveability.

Also, in spite of the fact that interventions by the administration's experts in public debates often referred to vague participatory actions, the programme contents, small in number and lacking in content, were the subject of criticism and challenges by several associations operating in the districts.

In this part of the city—used as containers of the most diverse items, created without any concern for the creation of necessary connections with the social fabric of inhabitants, and in return for the false intention of equipping them with special facilities—the desire to support the construction market and increase the land revenue of the surrounding area is well hidden, also to assist the demand for housing that comes from the creditworthy bracket of citizens.

In fact, after almost thirty years, of the three programmes, not only have neither the public nor the private interventions been carried out, but they have been the fly-wheel for the approval of variations for the building of shopping malls, which were to be built after the creation of those redeeming public works that the city council was not able to carry out.

This is the case of the variations authorised by the City Council to build the *Conca d'Oro* shopping mall in an area close to the *ZEN 2 insulae* and the *Torre Ingastone* shopping mall in the Borgo Nuovo district (Fig. 6). Variations that, in the case of *ZEN 2*, saw the shopping mall built by the Società Immobiliare Monte Mare S.p.A. owned by Maurizio Zamparini as a necessary alternative to the creation of services, infrastructures and public facilities in the district. More than three years after the shopping mall was opened, the planned redeeming public works (a swimming pool, a large public park around *Villa Raffo*, a green area for sports, a children's centre, an elderly persons' centre and a municipal centre) have not yet been delivered to the City Council. A similar destiny for the planned park area around *Torre Ingastone*, close to the shopping mall of the same name in *Borgo Nuovo*. Privatisation of urban space

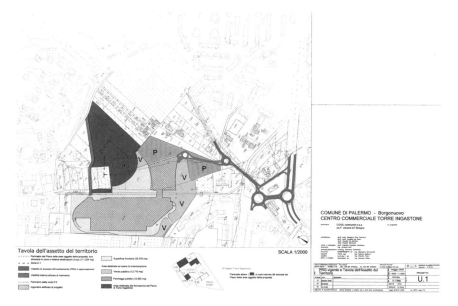

Fig. 6 Variation for the *Torre Ingastone* Shopping Mall on the 2004 *PRG* (Masterplan) (Source: *Comune di Palermo*, Variation of *PRG*, 2006)

continues and emphasises the long history of unmet needs that has characterised the physical-social reality of these districts, to the detriment of inhabitants taking on an active role in the city, and an urban marketing operation is not sufficient to hide the unhealthy distortions of the unfair system on which the *development* of Palermo is founded.

The construction of these shopping malls, that should have been the *price to pay* for the creation of public services and spaces, have actually brought about not only a clear reduction in the areas that can be allocated to public services and areas as foreseen by the 2004 *PRG*, but also the non-construction of most of the public interventions contained in the *ZEN*'s *PII* and Borgo Nuovo's *PRU*, subsequently included in the variations for the construction of shopping malls and which today have been included in the Three-Year Public Works Programme. This also shows the City Council's main interest in keeping the inhabitants of the *ERP* districts in a permanent condition of exclusion.

In the absence of a public policy for the more marginalised and more vulnerable brackets, the analysed inhabitants' self-organisation and self-management practices have operated in spite of the administration (Cellamare 2011), which responded at times by intervening coercively in the illegal and unlawful operations, other times deliberately ignoring the divestment process of their own prerogatives in managing and planning the territory, and other times supporting strong private interests through variations.

4 Conclusion

As illustrated in the previous paragraphs, in Italy—although not only in Italy—formal housing policies are clearly inadequate, being repressive or prone to privatise or abandon the public heritage; at the same time, self-organisation of public space is on the rise, and new ways and places of social production are spreading. The *inhabitants* of these spaces are sometimes very good at behaving in ways and forms that challenge the regulatory, control/repressive purposes of dominating groups (Paba 2003). In declining the multiple dimensions of the right to the city, Jordi Borja and Zaida Muxì (2003) identify the "right to illegality" as a part of it, intended as the political exercising of claiming fundamental rights through modes and forms outside relevant legal frameworks. Also, as part of this dimension of the right to the city, they also include recognition of the "city of citizenship", i.e. those urban spaces that rules define as illegal but that should be recognised and converted into cities through amendments to law.

Now it is worth wondering how research and self-help practices implemented by squatters can contribute to modify, both theoretically and practically, the legal system they originate from. Sandercock (2000) has rightly pointed out that reviewing the legal system and the laws stemming from it is a long-term objective, which requires extensive and stable lobbying and participation actions, during a time period that can involve even more than one generation. However, in spite of being a long and complex process, it is undoubtedly paramount to come up with widespread and recognised policies, regardless of changeable political positions. In this respect, the cumulative process of knowledge and experience—both in terms of self-help practices and drafting of local regulations on the use of public heritage—that we have been experiencing over the last few years in Palermo is not at all marginal or irrelevant. In the long run, these practices can be the seed of a substantial change in the legal foundations of the right of ownership of public goods, thus having a considerable impact on the housing policies through which the right to housing is substantially granted. If the right to the city can be re-interpreted as "a right to belonging to a place, whether in spaces that we call cities or do not" (Aalbers and Gibb 2014, 209), then taking the concepts to the extreme we could say that not having a house is tantamount to being deprived of the very right to urban life and urban spaces. Hence, public property is the element on which to rebuild urban welfare, which is progressively being eroded by securitisation policies. As stated by Harvey (2012), it is indeed through these emerging practices that the current values can be renovated and we can take on the challenges imposed by urban neo-liberalism to societies, also in terms of democracy and social and spatial justice.

If we consider the self-restoration practices of the public properties of Palermo as an alternative housing policy, we can observe that the active role of the homeless contributes to overcoming the stigmatisation which is implicit in formal housing policies based on the axiom homeless = need (moral need) while, at the same time, giving the homeless voice capability in social choices and public decision-making (Sen 2009). *Voice* not only as political protest (Hirschman 1970) but also as aspiring capability (Appadurai 2004), in other words contributing to develop a policy *by* the homeless rather than *for* the homeless. At the same time, the conversion of a public building, as previously highlighted, triggered in the homeless the desire for integration, both social (with the residents in the neighbourhood) and physical (with the rest of the city), by taking care of the community spaces in the building.

It is no coincidence that, in reference to the general strategy in the fight against social exclusion in the metropolitan city of Palermo, the presentation document of the multi-fund National Operational Programme for Metropolitan Cities 2014–2020 (*Programma Operativo Nazionale Città Metropolitane 2014–2020, PON METRO 2014–2020*) recognises that:

> one of the reference terms for the ability to self-define ones needs and self-manage the relative services (informally and not in a structured manner) by the local communities is the concept of insurgent city, as the ability to self-organise, to respond to its own unmet need or a need not adequately met by public action (i.e. District cleaning, gardening, self-managed social centres, urban vegetable gardens, etc.). In this sense, it can be considered that the activation of new services and social cooperation networks can also strengthen other inclusion actions supported by PON METRO. (Dipartimento per lo Sviluppo e la Coesione Economica 2014, 16)

However, aside from the formal recognition of the contribution of said practices, their actual size still today is not provided with answers, or with a political inter-locutor. As the introduction request in art. 72 of the 2018 Finance Act proves, the City

Council acknowledged the existence of a housing emergency, of which the illegal occupations are a macroscopic outcome, but was not able to propose a review of its own regulations on housing interventions, recognising the idea of cities and urban spaces subtended to such practices.

These practices, in fact, contain elements and proposals which are crucial to critically analyse the coercive and top-down policies enforced by institutions in order to guarantee a pre-set order—even a spatial one—rather than the general right to the city.

References

Aalbers MB, Gibb K (2014) Housing and the right to the city: introduction to the special issue. Int J Hous Policy 14(3):207–213

Appadurai A (2004) The capacity to aspire: culture and the terms of recognition. In: Rao V, Walton M (eds) Culture and public action. Stanford University Press, Stanford, pp 59–84

Bonafede G, Lo Piccolo F (2010) Participative planning processes in the absence of the (public) space of democracy. Plan Pract Res 25(3):353–375

Borja J, Muxì Z (2003) El espacio publico: ciudad y ciudadania. Electa, Barcelona

Camera di Commercio di Palermo, Istituto Tagliacarne (2013) Osservatorio Economico della provincia di Palermo. Il sistema socioeconomico della provincia tra recessione e potenzialità di sviluppo. Camera di Commercio, Palermo

Cellamare C (2011) Progettualità dell'agire urbano: processi e pratiche urbane. Carocci, Rome

Cremaschi M, Di Risio AP, Longo G, Lucciarini S (2007) Dinamiche territoriali e questione abitativa. In: Clementi A (ed) Reti e territori al futuro. Materiali per una visione. Ministero delle Infrastrutture e dei Trasporti, Rome

Dipartimento per lo Sviluppo e la Coesione Economica (2014) Programma Operativo Nazionale Città Metropolitane 2014–2020. Documento di Programma. MEF, Rome

Fregolent L, Torri R (2018) Introduzione. In: Fregolent L, Torri R (eds) L'Italia senza casa. Bisogni emergenti e politiche per l'abitare, FrancoAngeli, Milan, pp 11–15

Giampino A (2012) Ai margini delle politiche sociali. Il disagio abitativo tra nuovi contesti e nuovi soggetti. In: Pinzello I (ed) Verso una nuova politica della casa. Politiche pubbliche e modelli abitativi in Italia e in Spagna—Hacia una nueva política de vivienda social. Políticas públicas y modelos de vivienda en Italia y España. FrancoAngeli, Milan, pp 67–86

Giampino A, Lo Piccolo F (2016) Formal property rights in the face of the substantial right to housing. Public Sector 42:52–63

Harvey D (2012) Rebel Cities: From the rights to the city to the urban revolution. Verso, London

Hirschman AO (1970) Exit, voice, and loyalty: Responses to decline in firms, organizations, and states. Harvard University Press, Cambridge

ISTAT (2011) 8milaCENSUS. ISTAT, Rome

ISTAT (2017) La povertà in Italia. Anno 2017. ISTAT, Rome

ISTAT-Caritas (2014) Le persone senza dimora. ISTAT, Rome

Lo Piccolo F, Giampino A, Todaro V (2014) Palerme, ville sans domicile. Droit au logement: entre informalité et arrangements politiques. In: Maury Y (ed) Les coopératives d'habitants, des outils pour l'abondance. Repenser le logement abordable dans la cité du xxi siècle. Chairecoop, Lyon

Lo Piccolo F, Giampino A, Todaro V (2017) The housing emergency in Palermo between rights and legality. In: Gospodini A (ed) Proceedings of the International Conference on Changing Cities III: Spatial, Design, Landscape & Socio-economic Dimensions. Grafima Publications, Thessaloniki

Ministero dell'Interno (2017) Gli sfratti in Italia: andamento delle procedure di rilascio di immobili ad uso abitativo. Ministero dell'Interno, Rome

Ombuen S, Ricci M, Segnalini O (2000) I programmi complessi. Innovazione e Piano nell'Europa delle Regioni. Il Sole24Ore, Milan

Paba G (2003) Movimenti urbani. Pratiche di costruzione sociale della città, FrancoAngeli, Milan

Perin Cavallo M (2004) Riprendere la discussione sul problema casa: alcune questioni. Urbanistica Informazioni 194:30–31

Rigon A (2008) Le fondazioni bancarie per l'housing sociale. Urbanistica informazioni 221–222:35

Siciliana Regione (2015) Annuario statistico Regionale. Regione Siciliana, Palermo

Roma G (1994) Domanda marginale di abitazioni e politiche urbane. Urbanistica 102:16–17

Sampaolo S (2008) I nuovi temi della domanda abitativa. Urbanistica informazioni 221–222:36–37

Sandercock L (2000) When strangers become neighbours: Managing cities of difference. Planning Theory & Practice 1(1):13–30

Sbetti F (2008) Nuovi strumenti e nuovi attori per l'emergenza abitativa. Urbanistica informazioni 221–222:39–41

Sen A (2009) The idea of justice. Allen Lane, London

Tosi A (1994) Abitanti. Le nuove strategie dell'azione abitativa. ilMulino, Bologna

Touraine A (2017) Noi, soggetti umani. Il Saggiatore, Milan

New Inhabitants for the Re-use of Historical Territories in South-Eastern Sicily

Giuseppe Abbate

Abstract In spite of the multifaceted criticalities that characterise the Sicilian territory, South-Eastern Sicily appears to be of particular interest due to the active role that distinguishes it from the rest of the Island. The transformation processes taking place in South-Eastern Sicily, in fact, seem to be guided by new perspectives of sustainability, local development and good governance. The mix obtained from the extraordinary amount of tangible and intangible resources, from some favourable socio-cultural circumstances and from the creation of new transportation infrastructures, has contributed to promoting new forms of reuse of the historical building heritage by Italian and overseas investors who have purchased prestigious properties in the historic centres and in the agricultural areas of the Hyblaean hinterland, triggering unexpected processes of economic revitalisation.

1 Introduction

The set of development policies enacted in Italy from the 1950s onwards, to provide a solution to the *southern issue*, favoured assisted development that did not succeed in bridging the income gap with the rest of the Country and in reducing dependency on transfers from the national government (Cafiero 2000).

The worldwide financial crisis that has also affected Italy for over a decade and which is still a reality despite recent signs of recovery, has further weakened the southern area and cities. There has been a further drop in the GDP per capita, a huge increase in unemployment and a high level of emigration of educated individuals, mostly young ones. The constant loss of inhabitants is creating increasing numbers of unused buildings, with the consequent collapse of real estate market prices.

The cultural backwardness of the socio-economic and institutional *milieu* that has affected the rules of civil life, with the prevalence of patronage-like political relations

G. Abbate (✉)
University of Palermo, Palermo, Italy
e-mail: giuseppe.abbate@unipa.it

F. Lo Piccolo et al. (eds.), *Urban Regionalisation Processes*,
UNIPA Springer Series,
https://doi.org/10.1007/978-3-030-64469-7_6

121

that tend to favour particular groups rather than giving answers to collective needs (Trigilia 2012), has slowed down the development process and discouraged (local and external) investors in the South of Italy, even before the economic recession.

This situation has not allowed the challenges of globalisation to be addressed sufficiently by building development promotion or consolidation policies to make the southern areas and towns competitive.

In spite of the difficulties of the context, in some areas of the South of Italy and Sicily in particular, new progressive trends are appearing alongside regressive trends, new energy sources that power entrepreneurship that invests in the optimi-sation of tangible and intangible *internal resources*, the rediscovery of agriculture and the promotion of the available seaside/cultural tourist industry. The ongoing transformation processes in these areas seem to be guided by new prospects of local development and good governance as elements for the structuring of territorial opti-misation, at the same time being a guarantee of improved liveability and social and cultural sustainability of choices.

Using these considerations as a starting point, we chose the South-Eastern area of Sicily as the area for investigation, as it appears to be especially dynamic in relation to a more advanced development than other parts of the region. The area of study coincides with the part of the area that is named *Area dei rilievi del tavolato ibleo* in the Regional Landscape Territorial Plan Guidelines.[1] This context includes almost all the territory in the Provinces of Ragusa and Syracuse and falls within the so-called *Val di Noto*, one of the three large administrative areas that Sicily was split into from Norman times to 1812 (Fig. 1).[2]

2   The Territory's Identity Traits

The element that distinguishes the natural and anthropic traits of the Hyblaean area more than anything is the widespread presence of limestone that has been excavated, sculpted and carved into cave dwellings (Fig. 2), temples (Fig. 3), palaces (Fig. 4), churches (Figs. 5 and 6) and was also used in the large network of drystone walls that has become one of the main distinguishing identity traits of South-Eastern Sicily's agricultural landscape. The limestone has been carved over thousands of years by streams that gave rise to the so-called *cave*, the deep natural ravines through which the water runs. In the parts of these huge ravines where ancient settlements were built, the rocky crags have been built upon and characterised by a compact urban fabric, where roads become stairways in the steepest parts. The complex orography of the places has also allowed for a scenographic arrangement of churches and convents from the Baroque period. To aid road traffic, several stretches of the streams were covered in the second half of the twentieth century and many of the characteristic

---

[1] Assessorato dei Beni Culturali ed Ambientali e della Pubblica Istruzione (1999).

[2] The other two areas were *Val di Mazara* and *Val Demone*.

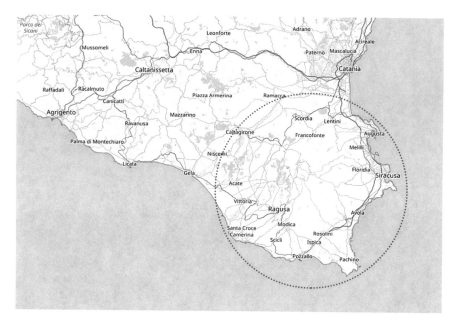

Fig. 1 The study area coinciding with the portion of territory which in the Landscape Regional Plan (*Piano Territoriale Paesistico Regionale, PTPR*) is called "Area dei rilievi del tavolato ibleo" (Image by Giuseppe Abbate)

Fig. 2 The Chiafura settlement (Scicli) with the cave dwellings (Photo by Giuseppe Abbate)

Fig. 3 Historic centre of Syracuse (Ortigia island): Temple of Apollo (Photo by Giuseppe Abbate)

stone bridges were therefore demolished, as in Modica and Scicli, thus erasing one of the elements that lent the urban landscape its identity.

The large-scale reconstruction carried out in the early decades of the eighteenth century, following the 1693 earthquake, brought new common traits to the towns in *Val di Noto*, expressed by the quantity and exceptional quality of the architectural and urban results in a late-Baroque style (Aymard 1985), further enriching the urban contexts that were already graced with the presence of fine architecture from ancient times that had survived the earthquake (Tables 1 and 2).

Starting from the second half of the eighteenth century, when Sicily began to be included in the destinations of those taking part in the *Grand Tour*, the *Val di Noto* became one of the favoured destinations for travellers, in awe of the beauty of the archaeological finds, especially from Classic Greek era, amidst a landscape of extraordinary naturalistic value. The South-Eastern city that carries the greatest attraction is Syracuse, due to its past as an Ancient Greek city. In particular, it was the complex of quarries in Syracuse, known as *latomie*, that captured the travellers' attention, which according to Brydone contained a large variety of wild, romantic aspects (Tuzet 1988).

When exploring the Hyblaean area, travellers are fascinated by the unusual rocky formations of the rock settlements of Pantalica and Cava d'Ispica, while they do not appear to like the *modern* towns that were rebuilt after the earthquake in 1693, that had *betrayed* the classic models (Trigilia 2006).

Fig. 4 Baroque palace in the historic centre of Syracuse (Ortigia Island) (Photo by Giuseppe Abbate)

Aside from the lack of appreciation for the Baroque towns, that are now, on the other hand, a priceless resources and world heritage site, travel diaries, drawings and guache paintings by travellers in the eighteenth century began to form the identity of an area, in the collective imagination, where the agricultural and natural characteristics, together with the thousands of years of history of settlements formed an exceptional, unique situation (Tables 3 and 4).

The landscapes so appreciated by the *Grand Tour* travellers are not contrasted by the recent, man-made landscapes that in the inland Hyblaean area are still environmentally integral while on the coast, they are often congested, deteriorated and polluted.

The most glaring case is that of the large area affected by the Syracuse petrochemical district, that lies in the area of Augusta, Priolo and Melilli, which is now suffering from problems of reconversion and lack of decontamination, after the heavy drop in industrial activities.

Several stretches of coastline in South-Eastern Sicily contain buildings only used in certain periods of the year, and structures that are mainly destined for mass tourism use (La Greca and La Rosa 2012), with a percentage of illegal constructions that is lower, however, than in other coastal areas of Sicily. The stretch of coastline between Syracuse and Santa Croce Camerina contains a number of land parcels, located along

Fig. 5 Scicli: the church of St. Bartolomeo (Photo by Giuseppe Abbate)

roads that are perpendicular to the coast, alternating with stretches of land covered by green houses, that have changed the traditional agricultural landscape.

Lastly, especially along the African-facing side, there are still some beautiful natural landscapes, where traces can still be found of the dune-system or humid areas that are now protected nature reserves.

Fig. 6 Modica: the church of St. Giorgio (Photo by Giuseppe Abbate)

Table 1 Architectures surveyed in the historic centres of the Ragusa Province

ISTAT code	Municipality	Surveyed architectures	Origins of historic centre
88009	Ragusa	161	B
88001	Acate	16	C–D
88002	Chiaramonte Gulfi	73	B–D
88003	Comiso	60	A–D
88004	Giarratana	29	D
88005	Ispica	58	B–D
88006	Modica	175	B–D
88007	Monterosso Almo	43	B–D
88008	Pozzallo	20	B–C
88010	Santa Croce Camerina	38	C
88011	Scicli	93	B–D
88012	Vittoria	135	C–D

Author's elaboration on ISTAT data

3 Current Trends

The flourishing agricultural industry with the production of greenhouse-grown products, which sustains a large part of the local economy, but also growing attention towards new forms of cultural and gastronomic tourism, make South-Eastern Sicily one of the most dynamic areas in the region.

In spite of the rhetoric characterising this area (the drystone wall landscape, traditional places, Baroque area), South-Eastern Sicily actually has a better quality of life compared to the rest of the region and its towns are in a good condition, an indicator of systematic attention to maintenance and therefore good urban governance (Cannarozzo 2010; Abbate 2015a).

The creation of large transportation infrastructures such as Comiso Airport, the tourist port at Marina di Ragusa and two new stretches of the A18 motorway, which will connect Syracuse to Gela when completed, have also encouraged an increase in tourist flows and Italian and overseas investors in South-Eastern Sicily in recent years. "Pio La Torre" Airport in Comiso, opened in 2013, makes new use of an airport structure built during the fascist era and then expanded in the 1980s when Comiso had a NATO base that was subsequently dismantled.

Today, the airport, desired by the local community, connects this part of Italy to several Italian and European cities.

The tourist port at Marina di Ragusa, opened in 2009, was built thanks to a *project financing* operation linked to European funding. This is a state-of-the-art facility with

Table 2 Architectures surveyed in the historic centres of the Syracuse Province

ISTAT code	Municipality	Surveyed architectures	Origins of historic centre
89017	Syracuse	222	A–B
89001	Augusta	43	B–D
89002	Avola	54	D
89003	Buccheri	15	B–D
89004	Buscemi	32	B–D
89005	Canicattini Bagni	19	C
89006	Carlentini	19	C–D
89007	Cassaro	23	C–D
89008	Ferla	32	D
89009	Floridia	20	C–D
89010	Francofonte	66	B–D
89011	Lentini	60	A–D
89012	Melilli	49	B–D
89013	Noto	83	D
89014	Pachino	16	C
89015	Palazzolo Acreide	165	A–B
89020	Porto Palo di Capo Passero	0	C
89021	Priolo Gargallo	0	C
89016	Rosolini	25	C
89018	Solarino	4	C
89019	Sortino	47	D

900 boat moorings and is one of the main landing points in Sicily for boats up to fifty metres long.

After years of inertia, two new stretches of the A18 motorway (Syracuse-Gela) were completed in 2008: the Cassibile-Noto stretch, about 14 km long, and the Noto-Rosolini stretch, about 16 km long. 2010 saw the start of work for the two new stretches of motorway with relative exits at Ispica and Modica.

In the case of Syracuse and Ragusa, the two largest cities in the study area, two specific legislative provisions issued by the Sicilian Regional Assembly (*Assemblea Regionale Siciliana, ARS*) played a decisive role in the rebirth of the respective historical old towns, namely RL 70/1976 "Protection of historical town centres and special regulations for the Ortigia district of Syracuse and for the historical old town of Agrigento", and RL 61/1981 "Regulations for the recovery and renovation of the historical old town of Ibla and some districts of Ragusa".

Table 3 Isolated architectures surveyed in the Ragusa Province

Municipality	Military architect.	Religious architect.	Residential architect.	Productive architect.	Total
Ragusa	7	5	21	16	49
Acate	3	0	0	6	9
Chiaramonte Gulfi	1	3	2	22	28
Comiso	0	2	5	9	16
Giarratana	0	1	0	10	11
Ispica	0	2	6	7	15
Modica	11	4	32	15	62
Monterosso Almo	0	2	0	13	15
Pozzallo	1	1	2	0	4
Santa Croce C.	1	1	1	2	5
Scicli	8	1	5	17	31
Vittoria	1	3	4	4	12

Cannarozzo (2001)

Although with several years of delay compared to the time established in RL 70/1976, Syracuse was the first Sicilian city to have, in 1990, a historical old town recovery plan (at the same time as the one for the district of Ortigia). The existence of this detailed plan for Ortigia was the basis on which over the years local councils had built their policies for the recovery and upgrading of the historical city, obtaining European funding in order to better support public and private recovery work (Abbate and Orlando 2014) (Fig. 7).

In spite of delays in the drawing up of a plan for the historical old town, as provided for by RL 61/1981, Ragusa also used large amounts of funding provided by the same law, allowing hundreds of properties to be purchased and renovations to be carried out on prestigious buildings, while awaiting the drafting of an urban planning tool (Trombino 2004), which was only completed and approved in 2014.

The municipal councils of Scicli (Figs. 8, 9, 10, and 11) and Modica (Figs. 12, 13, 14, and 15), in 2009 and 2014 respectively, set up preparation for their fact-finding and interpretation frameworks referring to the various traits of the *ancient city* that led to the writing of guidelines for the protection and optimisation of the respective historical centres, with identification of project strategies that must be developed into legislation, by the drafting of two PRG general Variants for the respective historical old towns.[3]

---

[3] The legislative reference for drawing up said general Variants is the *Circolare 3/2000* of the Sicilian Regional Administration Territorial and Environmental Department.

Table 4 Isolated architectures surveyed in the Syracuse Province

Municipality	Military architect.	Religious architect.	Residential architect.	Productive architect.	Total
Syracuse	9	7	17	133	166
Augusta	4	5	1	33	43
Avola	1	2	0	1	4
Buccheri	0	2	0	4	6
Buscemi	0	1	0	2	3
Canicattini Bagni	0	1	2	4	7
Carlentini	1	2	0	34	37
Cassaro	0	1	0	1	2
Ferla	0	1	0	1	2
Floridia	0	0	0	7	7
Francofonte	1	1	1	9	12
Lentini	0	1	0	54	55
Melilli	0	3	0	64	67
Noto	6	6	12	74	98
Pachino	2	2	0	7	11
Palazzolo Acreide	1	1	4	7	13
Porto Palo di Capo Passero	2	0	0	4	6
Priolo Gargallo	2	2	3	15	22
Rosolini	2	2	4	4	12
Solarino	0	1	0	6	7
Sortino	0	1	2	3	6

Cannarozzo (2001)

To draw up the guidelines, the two administrations decided to use scientific advice offered by the Interdepartmental Research Centre on Historical Towns (*Centro Inter-dipartimentale per la Ricerca sui Centri Storici, CIRCES*), a research department at the University of Palermo, which aims to provide consultancy services and scientific support to territorial body initiatives on urban and territorial upgrading and historical old town renovation policies and plans[4] (Abbate et al. 2010; Abbate 2015b; Trombino 2016).

The inclusion of *eight cities in South-Eastern Sicily* in the list of "UNESCO World Heritage Sites" contributed significantly to allowing an international public to discover South-Eastern Sicily.

---

[4]The author was a part of two different *CIRCES* work groups who drew up the guidelines for the protection and optimisation of Scicli's and Modica's historical old town centres.

Fig. 7 Historic centre of Syracuse (Ortigia island): an image of piazza Duomo (Photo by Giuseppe Abbate)

These cities were Caltagirone, Catania, Militello in Val di Catania, Modica, Noto, Palazzolo Acreide, Ragusa and Scicli, which provide an exceptional testimony of what was the peak and final flourish of Baroque art in Europe.

In 2005, "Syracuse and the Rocky Necropolis of Pantalica" were included in the list of UNESCO sites. Together they form an extraordinary testimony of the development of different Mediterranean cultures through three thousand years of history. Another extraordinary tourist promotion tool for the Ragusa area, part of the phenomenon known as *film-induced tourism,* was the fictional TV series *Il Commissario Montalbano,* broadcast from 1999, in Italy, but also in another nine European countries (Spain, France, United Kingdom, Denmark, Sweden, Finland, Germany, Poland and Hungary), in the United States and in Australia. The successful fiction, based on the novels written by the writer Andrea Camilleri, chose the Ragusa area as the location for almost all the episodes, despite the fact that the places described by the author in his novels were based on areas in the Agrigento area (Vigàta is Porto Empedocle; Montelusa is Agrigento), making the cities of Ragusa, Modica and Scicli, the coastal villages of Sampieri (Fig. 16) and Donnalucata (Figs. 17 and 18) and, more generally, the wonderful Hyblaean area landscape internationally famous.[5] It is therefore a combination of factors (historical-artistic, landscape,

---

[5]From the 1960 s onwards, several film directors have set their works in South-Eastern Sicily. Some of the best-known films were *Divorzio all'italiana* by Pietro Germi, *Gente di rispetto* by Luigi Zampa, *Kaos* by the Taviani brothers, and *L'uomo delle stelle* by Giuseppe Tornatore.

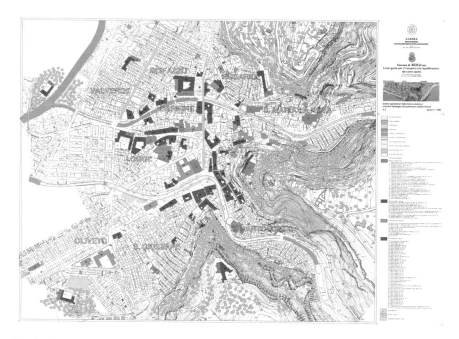

Fig. 8 Guidelines for the recovery and redevelopment of Scicli, Donnalucata and Sampieri historic centres. Table A.3—Scicli: System generating the urban structure and typological classification of the historical building heritage and the public spaces (Image by Giuseppe Abbate)

Fig. 9 An image of Scicli historic centre (Photo by Giuseppe Abbate)

Fig. 10 Scicli: the "Cava" of St. Maria La Nova (Photo by Giuseppe Abbate)

climate and relational qualities) that makes the context in question especially attractive, that influenced the recorded increase in tourist flows and a growing interest
from part of Italians from other regions and foreigners to invest in historical centres
and agricultural areas in the Hyblaean hinterland. These are people who belong to
a high-level socio-economic class in search of prestigious real estate properties to
transform into seasonal residences and, in a few cases, into permanent residences.

Starting from the fact that a large percentage of *rich* foreigners who invest in
second houses in South-Eastern Sicily tend to escape any traditional forms of census
due to the objective difficulty in retrieving statistical data, and can therefore be

Fig. 11 Beneventano palace in the historic centre of Scicli (Photo by Giuseppe Abbate)

estimated, for example via reports from real estate agencies operating in the area, only some basic considerations can be given about foreigners who have decided to move to this part of Sicily, based on some statistical data provided by *ISTAT*.

An analysis of the latter, in their evolution from 2011 to 2018, shows that the most significant increase in foreign presences in the Provinces of Ragusa and Syracuse, mostly concerns foreigners from "the South of the world", who meet the labour market demand in the agricultural sector, but also highlights the presence of small numbers in overall numerical terms, but which are still significant if compared to the local area, foreigners from Northern Europe, from Malta and from North America

Fig. 12 Guidelines for the recovery and redevelopment of Modica historical centre. System generating the urban structure and typological classification of the historical building heritage and the public spaces (Image by Giuseppe Abbate and Giuseppe Cascino)

Fig. 13 Modica: Grimaldi street (Photo by Giuseppe Abbate)

(United States and Canada), who are constantly increasing but with non-significant variations that are not comparable to the increases recorded each year in the total number of foreigners from other parts of the world (Tables 5 and 6).

Real estate investment is mostly aimed at the renovation of prestigious buildings, in the historical old town areas (Ragusa, Modica, Scicli, Noto, Ortigia) and in the Hyblaean hinterland.

Fig. 14 Vaulted passage in the historic centre of Modica (Photo by Giuseppe Abbate)

Fig. 15 An image of Modica historic centre (Photo by Giuseppe Abbate)

Fig. 16 An image of Sampieri historic centre (Photo by Giuseppe Abbate)

The intention is to recover historical buildings transforming them into summer houses and in a few cases, once the owners have retired, into permanent homes, but also into hotels, bed & breakfasts, holiday villas, residences and farmhouse guesthouses, usually of a very high standard (Figs. 19 and 20).[6]

---

[6]In the last decade in South-Eastern Sicily, the number of hotels has doubled, with a significant growth in four-star and five-star structures, while other types of tourist accommodation have tripled.

Fig. 17 Historic centre of Donnalucata: marina street (Photo by Giuseppe Abbate)

Fig. 18 An image of Donnalucata historic centre (Photo by Giuseppe Abbate)

The creation of the new tourist port of Marina di Ragusa, in particular, has brought opportunities to many boat-owning Maltese inhabitants to invest in the purchase of real estate in the agricultural areas of South-Eastern Sicily, who, considering the short distance from Malta (about 50 miles) are then converted into weekend houses or are transformed into flourishing farms, meeting the demand for agricultural land, that is now difficult to find on the island of Malta, by the Maltese community (Figs. 21 and 22).

Table 5 Overseas residents of European and North American origin in the Province of Ragusa

	2011	2012	2013	2014	2015	2016	2017	2018
Austria	14	14	14	10	12	14	13	12
Belgium	14	15	12	19	22	20	20	23
Denmark	2	2	2	3	3	4	4	2
France	39	45	48	61	66	73	69	75
Germany	118	122	121	132	133	141	146	144
Netherlands	12	13	14	14	13	14	16	22
United Kingdom	19	26	30	38	33	43	54	58
Sweden	7	7	6	5	5	8	9	9
Switzerland	6	6	9	8	12	14	14	15
Malta	25	27	23	23	26	26	29	28
United States	26	23	28	30	39	38	41	45
Canada	2	4	4	4	7	8	10	9
Total overseas	284	304	311	347	371	403	425	442

Author's elaboration on demo.istat.it data

Table 6 Overseas residents of European and North American origin in the Province of Syracuse

	2011	2012	2013	2014	2015	2016	2017	2018
Austria	22	26	28	23	22	20	23	21
Belgium	30	31	37	31	29	35	38	39
Denmark	2	2	2	2	2	2	4	4
Finland	14	17	18	17	14	13	7	7
France	90	94	118	99	122	128	129	134
Germany	164	171	187	164	164	162	165	189
Netherlands	24	26	27	31	29	29	30	34
United Kingdom	105	114	137	107	131	133	140	160
Sweden	6	7	6	6	10	10	10	11
Switzerland	20	21	24	21	23	26	26	31
Malta	26	20	29	27	31	33	35	36
United States	44	50	65	51	56	56	66	69
Canada	15	16	20	13	15	15	16	28
Total overseas	562	595	698	592	648	662	689	763

Author's elaboration on demo.istat.it data

Fig. 19 Modica: an example of historical building recovered (Photo by Giuseppe Abbate)

Fig. 20 Scicli: examples of historical buildings recovered (Photo by Giuseppe Abbate)

Fig. 21 Rural architecture in the Hyblaean area (Photo by Giuseppe Abbate)

Fig. 22 Rural architecture in the Hyblaean area with a typical garden (Photo by Giuseppe Abbate)

4   Conclusions

The presented framework shows a new component in the inhabitants of South-Eastern Sicily, that is traced to the presence of foreigners whose socio-economic character-istics and living requirements are totally different to those of the native residents or immigrants worker, who are increasing the demand for *leisure* housing, which was until recent years not even considered in the area contexts in question (Lo Piccolo et al. 2013).

The presence of new, educated and affluent inhabitants has in some ways contributed to mitigating the current real estate market crisis that has affected Sicily for several years, although to a different extent in different areas, helping to record a positive trend for the sale of prestigious historical real estate.

This phenomenon has also contributed to reversing the depopulation trend of Baroque towns and the agricultural areas of the Hyblaean hinterland, revitalising the local economy in a totally unexpected manner. The renovation work on historical buildings by new inhabitants is usually of a good quality, an indicator of refined clients who prefer to entrust their requests to local professionals and master tradesmen with proven experience.

As the same time as the increasing presence of new affluent inhabitants in the historical old towns and agricultural areas of South-Eastern Sicily, this phenomenon cannot be left to the mere free initiative of the real estate market.

The local councils involved must adopt suitable social rebalancing policies and also refine targeted actions for offering a wide range of services, and an overall increase in the quality of public spaces, not just through urban upgrading work but also in terms of managing open spaces, guaranteeing cleanliness and safety, which would surely help support the international tourist demand but also that of the inhabitants who have increasing awareness of their needs regarding local development.

References

Abbate G (2015a) Nuovi abitanti per il riuso dei territori storici: il caso dell'area sud-orientale della Sicilia. In: Atti della XVIII Conferenza Nazionale SIU, Società Italiana degli Urbanisti, Venice, 11–13 June 2015, Planum Publisher, Rome-Milan, pp 1365–1371

Abbate G (2015b) Processi di rigenerazione nei centri urbani della Sicilia sud-orientale. Urbanistica Informazioni 263:3–6

Abbate G, Cannarozzo T, Trombino G (2010) Centri storici e territorio. Il caso di Scicli – Historical towns and their hinterland. The Scicli case study. Alinea, Florence

Abbate G, Orlando M (2014) Tutela dei centri storici e norme speciali per Siracusa e Agrigento. In: Iacomoni A (ed) Questioni sul recupero della città storica. Aracne, Rome, pp 137–149

Assessorato dei Beni Culturali ed Ambientali e della Pubblica Istruzione (1999) Linee Guida del Piano Territoriale Paesistico Regionale, Regione Siciliana, Palermo

Aymard M (1985) La città di nuova fondazione. In: De Seta C (ed) Storia d'Italia. Insediamenti e territorio. Einaudi, Turin, pp 405–414

Cafiero S (2000) Storia dell'intervento straordinario nel Mezzogiorno. 1950–1993. Lacaita, Manduria-Bari-Rome

Cannarozzo T (ed) (2001) Il Sistema dei centri storici. Assessorato Territorio e Ambiente della Regione Siciliana and Dipartimento Città e Territorio dell'Università degli Studi di Palermo, Palermo (*mimeo*)

Cannarozzo T (2010) Il Comune di Scicli nel contesto territoriale. In: Abbate G, Cannarozzo T, Trombino G, Centri storici e territorio. Il caso di Scicli - Historical towns and their hinterland. The Scicli case study. Alinea, Florence, pp 37–38

La Greca P, La Rosa D (2012) Val di Noto. Stanzialità turistica e trame insediative. In: Leone NG (ed) Itatour. Visioni territoriali e nuove mobilità. Progetti integrati per il turismo nell'ambiente. FrancoAngeli, Milan, pp 209–2019

Lo Piccolo F, Leone D, Lo Bocchiaro G (2013) La questione abitativa dei nuovi cittadini in Sicilia tra risposte emergenziali, lavori temporanei e nuove attrattività del territorio. In: Lo Piccolo F (ed) Nuovi abitanti e diritto alla città. Un viaggio in Italia. Altralinea, Florence, pp 103–129

Trigilia C (2012) Non c'è Nord senza Sud. il Mulino, Bologna

Trigilia L (2006) Siracusa, il Val di Noto e le città siciliane del Grand Tour. Annali del Barocco in Sicilia 8:87–95

Trombino G (2004) Ragusa, una città, due centri storici. L'Universo 4:436–457

Trombino G (ed) (2016) Modica. Contributi per il recupero e la riqualificazione del centro storico. 40due Edizioni, Palermo

Tuzet H (1988) Viaggiatori stranieri in Sicilia nel XVIII secolo. Sellerio, Palermo

Landscape of Exception as Spatial and Social Interaction Between High-Quality Agricultural Production and Immigrant Labour Exploitation

Vincenzo Todaro and Francesco Lo Piccolo

Abstract This Chapter analyses the spatial and social interaction phenomena between high-quality agricultural production and immigrant labour exploitation that produce the *landscape of exception*, a particular declination of the Agambenian (2005) "state of exception" concept. The *landscape of exception* construction mechanism is generated within South-Eastern Sicily through the productive system of greenhouses, finalised to the vegetables production. Greenhouses, in particular, represent an effective tool for spatial manipulation over the landscape and social control of migrant workers. In relation to these considerations, this work reflects on ethical challenges and dilemmas of planning, highlighting (both explicit and latent) conflicts and power inequalities in the *landscapes of exception*, where issues of social justice, environmental sustainability and suspension of norms are strictly intertwined.

1 Global North, Rural Regions and Immigration

In the northern hemisphere's rural regions, the non-specialised immigrant manual labour, mostly employed illegally and to excess, is an essential component for agricultural production systems. It provides a response to the demand for labour from the agricultural world that is now neglected by local populations (Ambrosini 2001; Cicerchia and Pallara 2009; Centro studi e Ricerche IDOS 2016; CoReRAS 2017). Therefore, even in the globalised rural world "immigrants' poor work is seen to be

---

[1] According to the ISMEA-SVIMEZ Report (2016), 70% of high-quality agricultural produce is concentrated in Southern Italy, and in particular in Campania, Calabria, Apulia and Sicily.

V. Todaro (✉) · F. Lo Piccolo
University of Palermo, Palermo, Italy
e-mail: vincenzo.todaro@unipa.it

F. Lo Piccolo
e-mail: francesco.lopiccolo@unipa.it

© The Editor(s) (if applicable) and The Author(s), under exclusive license to Springer Nature Switzerland AG 2021
F. Lo Piccolo et al. (eds.), *Urban Regionalisation Processes*,
UNIPA Springer Series,
https://doi.org/10.1007/978-3-030-64469-7_7

closely connected to and structurally necessary for rich work" (Ambrosini 2005, 60); it has become a key component of the neo-liberal rural economy.

However, the profile of excellence that often characterises some high-quality agricultural productions[1] does not correspond in any way to the living and working conditions of the immigrants employed in the agricultural sector.

This condition continues to be a serious critical factor in some geographically marginal and often fragile areas, also from an administrative-institutional point of view, where laws and regulations regarding employment and territorial management and planning are *suspended* by a perverse union between economic power and political power.

In these cases, economic power, the main source of income and development, manages to control the political sphere, *freezing* and not applying the regulations that are the basis of territorial and worker protection (especially immigrants), but which could damage it.

Moving away from theoretical considerations about the relations between forms of power and planning (Lo Piccolo and Halawani 2014; Lo Piccolo and Todaro 2015, 2018; Lo Piccolo 2016; Lo Piccolo et al. 2017a, 2017b; Todaro 2017), this Chapter intends to reflect on the matters concerning the relations between (economic) power and the transformation of the landscape. These relations define a *landscape of exception* intended as the spatial application of that "state of exception" that, according to Agamben (2005), is the outcome of the suspension—albeit legalised—of rights and regulations valid for everyone.

In particular, the document intends to analyse the cause-effect ratio between agricultural sector development, transformation of the landscape and immigrant workforce exploitation in the Province of Ragusa, highlighting how the configuration (and *planning*) of the landscape can create types of exploitation and social exclusion.

2 Migratory Flows and Local Economy[2]

Towards the end of the 1970s, a concurrence of factors, mainly traceable to the progressive adoption of restrictive immigration policies in Northern European countries, directed a large part of migratory flows traditionally heading to Northern Europe towards Southern Europe instead.

In this context, due to the concurrence of specific conditions, including a certain level of legislative and institutional tolerance of immigration, high levels of informality in entering the job market and the presence of a constant demand for labour in the primary sector and personal services (Caruso and Corrado 2012), there was a consolidation of what is commonly known as the "Mediterranean migration model" (Baldwin-Edwards and Arango 1999; King 2000). This model can be traced in particular, to the intensive agricultural production development system involving mono-functional specialisation.

---

[2]Data and considerations of this section do not take into account the effects connected to the Covid-19 crisis.

Similarly to what is happening in California with the exploitation of Mexican labour in the American agricultural sector (Martin 1985, 2002), Berlan (2001) refers to that "Californian model", the success of which is permitted by the immigrant population, due to its non-regulated, flexible, excessive and ethnically fragmented nature (Corrado 2012).

These conditions, which are now common to all Euro-Mediterranean countries (De Zulueta 2003; Baganha and Fonseca 2004), in particular to Greece (Kasimis et al. 2003; Kasimis and Papadopoulos 2005; Lambrianidis and Sykas 2009; Kasimis 2010), Spain (Hoggart and Mendoza 1999; Mendoza 2001; García Torrente 2002) and Italy (Reyneri 2007), define the profile of the area where working illegality and exploitation, as well as deprivation of the guarantee of rights, are indicative of a *distorted economic development* where foreign workers are both victims and instruments (Ambrosini 2015).

In Italy, although the final destinations of migratory paths are still mostly Northern regions rather than Southern, in the last thirty years, Sicily has changed from being a place of emigration to a place of reception for immigrants. It has, in fact, become the entrance gate to Europe, especially for flows coming from North Africa.

Since the 1970s, the island has therefore been one of the first regions in the South of Italy to be involved in international migratory flows, while natives are leaving the area, to emigrate abroad or to move to the industrial areas of the North.

Some of the most significant foreign immigrant worker settlements in Italy are on Sicily's coasts, and in particular close to the coasts of North Africa, such as Mazara del Vallo (in the Province of Trapani) and other intensive agriculture areas in the Provinces of Trapani and Ragusa, which host initial migratory flows, mostly Tunisian males (Zanfrini 1993; Pugliese 2006) (Fig. 1).

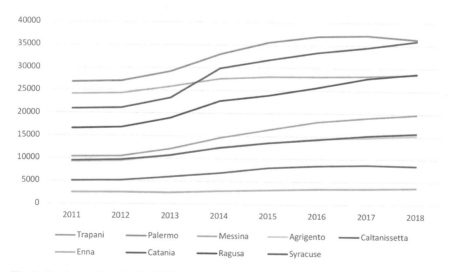

Fig. 1 Foreign residents in the Sicilian provinces on January 1 (Image by Vincenzo Todaro; data taken from *ISTAT* Census)

Compared to the first foreign presences in the 1970s, the new millennium has seen a diversification of migratory flows (Vertovec 2007) (Figs. 2–3). Generally, the following can be found:

- A migratory flow in transit (staying in the Southern regions while waiting to travel to the Centre and North);
- A second flow of legal and illegal workers who move from one district, usually agricultural, to another, following production cycles;
- A further flow of "retrogression migration" (Caruso and Corrado 2012) or *reverse mobility* compared to the traditional South/Centre-North direction, that sees a growing number of immigrants who were once employed in the manufacturing and constructions sectors in the North and Centre of the country moving to low-qualification service sectors (in cities) and to agriculture (in rural contexts) in the South (Colucci and Gallo 2015).

Other recent, significant changes to the initial migratory model have also been recorded in relation to the working sector. Although the agricultural sector has traditionally intercepted most of the immigrant labour offer, it is now becoming one of the various sectors of employment, and not the only one.

In fact, other equally unstable, temporary jobs, such as home caregiver, house cleaning, tourist-hotel positions and building work are also increasingly present.

Fig. 2 First aid and reception centre in Pozzallo (Ragusa) (Photo by Medici Senza Frontiere)

Fig. 3 First aid and reception centre in Pozzallo (Ragusa), interior space (Photo by Medici Senza Frontiere)

Another new element was introduced with the consolidation of flows from Eastern Europe, following the expansion of the EU in 2007.

This brought about further change, in particular in the agricultural sector: in those contexts where agricultural production is characterised by intense labour exploitation, European Union workers are preferred to non-EU workers (Ambrosini 2015), in order to avoid the crime of aiding and abetting illegal immigration.

This is also non-specialised, low-cost labour, mostly illegal and to excess (thus badly paid and not protected) that gives rise to critical issues in terms of social, economic and healthcare matters.

Lastly, the crisis in the agricultural sector, that in Sicily continues to be the *first absorption sector* for immigration, produces negative effects, especially on lowering of foreign workers wages and work demand that comes almost exclusively from low-level working environments (Centro studi e Ricerche IDOS 2016).

3 The Case of South-Eastern Sicily

With regard to the national agricultural policy framework, South-Eastern Sicily (Fig. 4) has traditionally occupied a prime role (ISMEA-SVIMEZ 2016), with its agricultural sector being significantly transformed in recent decades thanks to entrepreneurial ability and production, with significant international export figures (Asmundo et al. 2011).

Fig. 4 The plain of Vittoria (Ragusa) seen from the Hyblaean Mountains (Photo by Vincenzo Todaro)

A significant input to the agricultural sector has been provided by the greenhouse production of flowers, fruit and vegetables, in addition to the already-renowned, high-quality wine production (Fig. 5).[3]

In particular, the Hyblaean vegetable district, concentrated along the "transformed strip" (Fig. 6) of the Province of Ragusa (including the towns of Vittoria, Acate, Ispica, Scicli, Pozzallo, Comiso and Santa Croce Camerina) is the heart of an economic system with about 9000 companies with about 26,000 workers, that manage just over 9000 hectares of Utilised Agricultural Area (UAA), 2/3 of which is dedicated to greenhouse fruit and vegetable cultivation (Infocamere 2017).

The introduction of greenhouses in South-Eastern Sicily dates back to the end of the 1950s, when some farmers decided to convert their fruit and vegetable production from open-field production to greenhouse production.

The rapid success of greenhouse agriculture is linked to the triplication of annual production figures compared to open-field production. In a short time, this has led to the spread of this production system around the area, wherever possible, but above all where there were unused areas considered to be *non-productive*, especially along the coast.

---

[3]One of the main, high-quality productions is that of wine, including *IGP* (*Indicazione Geografica Protetta*) and *DOC* (*Denominazione di Origine Controllata*) wines. The leading wine is *Cerasuolo di Vittoria*, which holds an important place in the Ragusa Province economy.

Fig. 5 The tomato production into greenhouses of the Ragusa coast (Photo by Vincenzo Todaro)

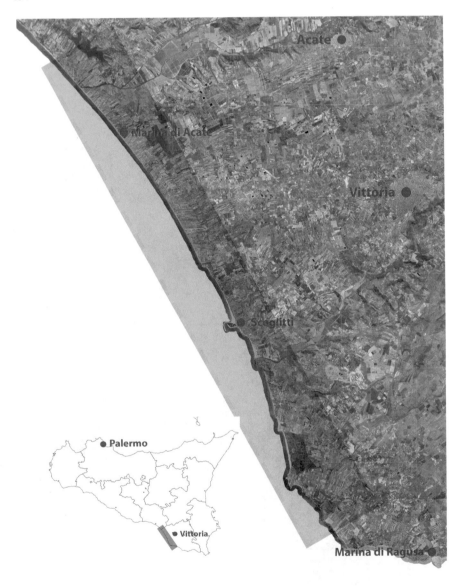

Fig. 6 The "transformed" strip along the Ragusa coast (Image by Vincenzo Todaro, https://www.google.it/maps/)

The first spatial effect was the disappearance of the sand dune system (*macconi*), in particular along the stretch of coast between Marina di Acate (Acate) and Punta Secca (Santa Croce Camerina), where over time, the natural modelling phenomenon had created a landscape of high naturalistic value, interspersed with humid sunken areas and characteristic dune vegetation (Campione 1994).

Fig. 7 The expanse of greenhouse plastic along the "transformed strip" (https://www.google.it/maps/)

In fact, it was necessary to level the ground to build the greenhouses, flattening the dunes which were thus reduced to small fragments amidst the expanse of greenhouse plastic (Figs. 7–8). The final effect is a huge transparent surface (due to the reflection of light on the flat surface of the plastic) that extends evenly and continuously along the entire area up to the sea, levelling out the natural depressions and mounds in the ground (Campione 1994).

In this way, the entire local economy has been consolidated around greenhouse production over recent decades, initially as family-run businesses, then later as industrial enterprises, at least for large companies.

The new territorial layout created by the greenhouses forms a *landscape of exception* that, referring to the "state of exception" by Agamben (2005), intended as the result of a suspension of rights and regulations, is a possible spatial variation of this phenomenon.

In international literature on the "state of exception", the condition of *exception* is mainly linked to the relationship between *strong powers* and *weak powers*, in particular in cases that are much publicised in the media, such as situations of war (Armitage 2002; Ek 2006; Gregory 2006), for example. In our case, the condition of *exception* witnessed through the suspension of environmental and urban-planning

Fig. 8 The expanse of greenhouse plastic along the "transformed strip" (Photo by Vincenzo Todaro)

regulations and laws is more silent and seemingly *invisible*, taking place in an apparently marginal context through precise spatial configuration (*space of exception*), that of the *greenhouse landscape*.

In this sense, the nearest references would be the ones relating to the spatial consequences of the "state of exception" (Baptista 2013; Gray and Porter 2014; Lo Piccolo and Halawani 2014; Lo Piccolo and Todaro 2015, 2018). Specifically, the *space of exception* of the greenhouse landscape in the province of Ragusa manifests itself in two ways: spatial and social. The former through the suspension of regulations and forecasts contained in plans, while the latter via the suspension of immigrant workers' rights.

4   *Landscape of Exception*: **The Spatial and Social Dimension**

The spatial dimension of the *landscape of exception* is shown through the suspension laws and forecasts contained in the plans. With regard to plans, the province of Ragusa is the part of Sicily with the most plans, for both urban-territorial planning (Provincial Territorial Plan, Hyblaean Land Strategic Plan, municipal urban-planning plans), and for environmental and landscape protection ("Pino D'Aleppo" Reserve Organisation Plan, "Vallata del Fiume Ippari" European Union Site of Importance Management Plan, Landscape Plan).

The strategic content of these plans, regarding the greenhouse issue, shows full awareness of the impact created on the environment and landscape by the greenhouses and, regarding the varying purposes of these tools, they agree on the need for them to be relocated and for the "transformed strip" to be completely reconverted and recovered.[4]

By comparing the forecasts contained in these plans and the territorial reality, however, some doubts arise about the efficacy of the plans. The main issue concerns the great inconsistency between the plans' forecasts and the real situation in loco.

The reasons behind this condition can be understood through the suspension (and not application) of the forecasts contained in the said plans. In fact, although the plans provide for the dismantling of the greenhouses, their implementation is postponed in time to avoid any considerable economic damage that such action may cause.[5]

According to these considerations, the economic power that a greenhouse production system has can place heavy pressure on political and administrative power. The latter should exercise their rightful forms of control over implementation of the plans, but, in this case, *suspends* their application, deferring them.

With regard to the spatial point of view, the *greenhouse landscape* has substituted the traditional coastal dune landscape, while from a social point of view the greenhouses have produced a thorough transformation of the *social landscape* of the area (Todaro 2015, 2016, 2017).

This has given rise to a "differentiated rurality" (Corrado 2012), the result of the change in the ethnic and economic composition of the local community, but at the same time a structural part of the greenhouse agricultural production development model (Berlan 2008; Colloca and Corrado 2013). In regard to this interpretation, the *social weakness* of immigrants is an essential part of this *landscape of exception*.

In relation to the "transformed strip" of the province of Ragusa, in fact, non-official statistics (Caritas Migrantes 2011; INEA 2013) show a significant concentration of foreign workers[6] employed in the greenhouses who help to triple the annual production of fruit and vegetables.

However, there is an extremely complex reality behind this economic success, that sees immigrants living in severely distressing working and living conditions (Medici Senza Frontiere 2007).

---

[4]In particular, the Landscape Plan (2007), drafted by the Ragusa Cultural and Environmental Heritage Department, studies the greenhouse issue using a specific tool, the "Macconi" Environment Project. This project divides the greenhouse area into three zones (red zone on the beach, recovery zone within 150 metres of the sea, background recovery zone) for which it specifies the methods for reconverting and environmentally recovering it.

[5]To have an idea of the economic value of the greenhouse productions, 1 hectare of cultivated vegetable growing surface area in a greenhouse can correspond to the production of more than 70 hectares of arable land, although the management cost of a greenhouse farm is three times that of a crop-growing farm (ISTAT 2016).

[6]According to *ISTAT* figures ("Population and dwelling census") for the decade 2001–2011, the foreign population in Sicily more than doubled in that time, from 49,399 to 125,015 units. The top five foreign nationalities in Sicily—Romanian, Tunisian, Moroccan, Sri Lankan and Albanian—represent over 50% of the entire foreign presence. In relation to the 2011 figure, the Province of Ragusa has the highest number of foreign presences compared to the resident population.

Fig. 9 Migrant workers house near the greenhouses of the "transformed strip" (Photo by Vincenzo Todaro)

If we observe the living conditions found in these rural areas, we can see an extremely serious situation: the immigrants often live in small abandoned, precarious buildings found near the fields or greenhouses, and therefore a long distance from towns and facilities (Osti 2010) (Fig. 9).

Work in the greenhouses, the number of hours worked, the distance from the towns and the lack of transport that makes them *autonomous* are clear limits to the individual and collective freedom.

Other critical factors regard the lack of legal employment contracts that immigrants working in the greenhouses often endure. In this case, it is partly *black-market work* and partly *grey work* (not all declared).

The former, according to some cautious estimations of Caritas Migrantes (2011), affects 10% (and in some periods of the year, 20%) of the total of legally employed workers.

On the other hand, INEA (2013) estimates the overall number of immigrant agricultural workers to be 15,000–20,000 units and of these, estimates that 50–60% are illegally employed.

So-called "grey work", on the other hand, is where there is a legal employment contract (that allows the non-EU immigrant to obtains his permit of stay in Italy), by which the immigrant declares that he works approximately 102 days per year (the minimum to be able to accrue entitlement to unemployment benefit for the months in

Fig. 10 Migrant workers along the roads of the "transformed strip" (Photo by Vincenzo Todaro)

which he does not officially work), while actually he works the entire year (Figs. 10, 11 and 12).

In addition to this, another two phenomena denounce the serious conditions of immigrant greenhouse workers. The first regards damage to health (in particular dermatitis, gastroenteritis, respiratory problems, inflammation of the eyes) caused by continuous exposure over time to chemical products (pesticides and phytosanitary products) used in greenhouse productions to protect the produce, in particular from some kinds of fungus (INEA 2013).

Greenhouse products is, in fact, the agricultural field with the highest use of phytosanitary products. The province of Ragusa is the leading province in Sicily, and one of the leading provinces in Italy, for use of phytosanitary products in agriculture. In this area, in the ten-year period 2003–2013, said use increased significantly, with peaks in 2008 (8,407,301 tonnes) and in 2010 (8,263,907 tonnes) (INEA 2013).

Lastly, the latest serious phenomenon concerns the sexual exploitation of female immigrants. As reported by operators in the field (Caritas, Medecins sans Frontieres), this is a widespread phenomenon that involves female immigrant workers (mostly Romanian women), who are blackmailed by the greenhouse owners in order to keep their jobs. In this sense, the abnormal increase in the rate of abortions found in local provincial healthcare facilities from 2004 significantly affects immigrant female workers from Eastern Europe (Scucces Ferraro 2006).

In connection with this specific phenomenon, the structural characteristics of the *greenhouse landscape* guarantee *invisibility* (Le Blanc 2009), thus allowing the

Fig. 11 Migrant workers awaiting recruitment at the beginning of the working day in the district of Cassibile (Syracuse) (Photo by Vincenzo Todaro)

Fig. 12 Presence and labour exploitation of children in the greenhouses (Vittoria) (Photo by Vincenzo Todaro)

illegal workers to be unseen from the outside and also hiding the various forms of abuse and exploitation that they fall victim to.

Considering the extent and gravity of these phenomena that cause various types of exploitation to be carried out on the immigrants, however, the institutional players and official policies (including the ones contained in the planning tools) are completely absent (even for the institutional tasks that they should fulfill). The only operators in the area are voluntary associations and healthcare facilities that mainly respond to emergency situations due to the lack of available resources (Todaro 2014).

In reference to the controls carried out by the police forces, Ambrosini (2015) reveals a high level of tolerance in Italian regions, characterised by significant agricultural production compared to illegal work, that is instrumental for the correct functioning of entire local economies. On this matter, referring to the Ministry of Labour and Social Policies data (Ministero del Lavoro e delle Politiche sociali 2012), Ambrosini considers it to be a perfect example that in 2001, controls only found 361 workers in the whole of the South of Italy without a permit of stay. To all effects, this condition comes under the *status of exception* regulations suspension mechanisms.

5 Conclusions

It is a shared opinion that "immigration into rural societies in Southern Italy becomes a part of the functional and development methods of the South of Italy's social and economic system, giving new life […] consolidated relations of local power" (Mignella Calvosa 2013, 10).

The immediate consequence of this condition is generally the suspension of current laws, including the forecasts for planning tools. Their application is not formally cancelled, but they are merely not implemented or postponed in time. In the case of the province of Ragusa, what makes all this possible is the economic success of agricultural greenhouse production. In fact, in recent decades, this type of production has produced widespread forms of social redemption, in addition to distributing wealth throughout the area.

In addition to being the engine for this socio-economic development models, the greenhouses are also an efficient control and spatial and social handling tool. They are a decisive factor in the transformation of the traditional landscape, carrying out extraordinary violence on a natural ecosystem, including coastal dunes, which are commonly thought to be non-productive, but are of a great environmental and landscape value. This ecosystem, recognised locally, nationally and internationally, with the establishment of protected areas (Nature 2000 reserves and sites, of the "Habitats" EU Directive), has been reduced to isolated fragments and shreds. The protection system defined by laws and planning tools for this type of natural heritage has been suspended and is not applied, allowing the greenhouses to spread wherever possible, right up to the coastline.

From a social point of view, the greenhouses are a huge employment basin, a certain offer of work (unqualified and *flexible*) that can guide international immigration flows, who consider it to be an improvement on the living and working conditions in their country of origin (Kasimis 2010). However, suitable working conditions are not provided given the high level of job availability. Work in the greenhouses is extremely harsh, and in many cases, increasingly unsustainable (Avallone 2011). Citizens' rights, including legal contracts and fair pay, are denied and many forms of violence are carried out on the immigrants. Although the economic success of these agricultural productions is highly dependent on the immigrants' labour, they are not aware of this. However, the economic-production sector and also the political-institutional level are perfectly aware of this fact. The latter, on the one hand, is indifferent to the social policies applied to receiving the immigrants they should take responsibility for, and on the other hand, makes the territorial protection tools, including planning, ineffective. This brings into play the ethical side of planning, as, in various realms of intervention, this should decisively contribute to guaranteeing citizens' rights, via spatial and social balance (Lo Piccolo 2013).

References

Agamben G (2005) State of exception. University of Chicago Press, Chicago

Ambrosini M (2001) La fatica di integrarsi. il Mulino, Bologna

Ambrosini M (2005) Sociologia delle migrazioni. il Mulino, Bologna

Ambrosini M (2015) L'inserimento degli immigrati sul territorio. In: L'Italia e le sue regioni. Treccani, Rome. http://www.treccani.it/enciclopedia/l-inserimento-degli-immigrati-sul-territorio_%28L%27Italia-e-le-sue-Regioni%29/. Accessed 21 May 2019

Armitage J (2002) State of emergency. Theory, Culture and Society 19(4):27–38

Asmundo A, Asso PF, Pitti G (2011) Innovare in Sicilia durante la crisi: un aggiornamento di Remare controcorrente. StrumentiRes 4:1–7

Avallone G (2011) Sostenibilità, agricoltura e migrazioni. Il caso dei lavoratori immigrati nell'agricoltura del sud d'Italia. Culture della sostenibilità 8:1–12

Baganha MI, Fonseca ML (eds) (2004) New Waves: Migration from Eastern Europe to Southern Europe. Luso-American Foundation, Lisbon

Baldwin-Edwards M, Arango J (1999) Immigrants and the informal economy in Southern Europe. Frank Cass Pub, London

Baptista I (2013) Practices of exception in urban governance: Reconfiguring power inside the state. Urban studies 50(1):39–54

Berlan JP (2001) La longue histoire du modèle californien. Forum Civique Européen, El Ejido, terre de non-droit. Golias, Paris, pp 15–22

Berlan JP (2008) L'immigré agricole comme modèle sociétal? Études rurales 182:219–226

Campione G (1994) Sicilia. I luoghi e gli uomini. Gangemi, Rome

Caritas Migrantes (2011) Dossier statistico immigrazione 2011. XXI Rapporto. IDOS, Rome

Caruso FS, Corrado A (2012) Crisi e migrazioni nel Mediterraneo. I casi del Poniente Almeriense e della Piana di Sibari. Agriregionieuropa 31:58–60

Centro studi e Ricerche IDOS (2016) Dossier Statistico Immigrazione 2016. IDOS, Rome

Cicerchia M, Pallara P (eds) (2009) Gli immigrati nell'agricoltura italiana. INEA, Rome

Colucci M, Gallo S (2015) Tempo di cambiare. Rapporto 2015 sulle migrazioni interne in Italia. Donzelli, Rome

Consorzio Regionale per la Ricerca Applicata e la Sperimentazione (CoReRAS) (2017). Report sulle filiere agroalimentari siciliane. Antipodes, Palermo

Corrado A (2012) Ruralità differenziate e migrazioni nel Sud Italia. Agriregionieuropa 28(8):72–75

De Zulueta T (2003) Migrants in irregular employment in the agricultural sector of Southern European countries. Report for the Debate in the Standing Committee, Council of Europe

Ek R (2006) Giorgio Agamben and the spatialities of the camp. Geografiska Annaler B 88(4):363–386

García Torrente R (2002) La inmigración y el modelo de desarrollo almeriense II: Análisis de las necesidades de mano de obra en la economía almeriense. In: Pimentel M (ed) Mediterráneo Económico: Procesos migratorios. Economía y personas. Instituto de Estudios de Cajamar, Almería, pp 389–409

Gray N, Porter L (2014) By any means necessary: Urban regeneration and the "state of exception" in Glasgow's Commonwealth Games 2014. Antipode 47(2):380–400

Gregory D (2006) The black flag: Guantánamo Bay and the space of exception. Geografiska Annaler B 88(4):405–427

Hoggart K, Mendoza C (1999) African immigrants workers in Spanish agriculture. Sociologìa Ruralis 37(4):538–562

INEA (2013) Indagine sull'impiego degli immigrati in agricoltura in Italia 2011. INEA, Rome

Infocamere (2017) Movimprese 2017. https://www.infocamere.it/movimprese? Accessed 15 September 2018

ISMEA-SVIMEZ (2016) Rapporto sull'agricoltura del Mezzogiorno. ISMEA, Rome

ISTAT (2016) Le statistiche sull'agricoltura siciliana: informazioni per l'analisi delle politiche. Censimento dell'agricoltura 2010, vol 2. Leima, Palermo

Kasimis C (2010) Trend demografici e flussi migratori internazionali nell'Europa rurale. Agriregionieuropa 21:71–74

Kasimis C, Papadopoulos AG (2005) The multifunctional role of migrants in Greek countryside: Implications for rural economy and society. Journal of Ethnic and Migration Studies 31(1):99–127

Kasimis C, Papadopoulos AG, Zacopoulou E (2003) Migrants in rural Greece. Sociologia Ruralis 43(2):167–184

King R (2000) Southern Europe in the changing global map of migration. In: King R, Lazaridis GM, Tsardanidis C (eds) Eldorado or Fortress? Migration in Southern Europe. MacMillan, Basingstoke, pp 1–26

Lambrianidis L, Sykas T (2009) Migrants, economic mobility and socioeconomic change in rural areas: The case of Greece. European Urban and Regional Studies 16(3):237–256

Le Blanc G (2009) L'invisibilité sociale. PUF, Paris

Lo Piccolo F (2013) Nuovi abitanti e diritto alla città: riposizionamenti teorici e responsabilità operative della disciplina urbanistica. In: Lo Piccolo F (ed) Nuovi abitanti e diritto alla città. Un viaggio in Italia. Altralinea, Florence, pp 15–32

Lo Piccolo F (2016) Landing through informal blue infrastructures: The state of exception in planning. In: Moccia FD, Sepe M (eds) Reti e infrastrutture dei territori contemporanei. Networks and infrastructures of contemporary territories. INU Edizioni, Rome, pp 41–51

Lo Piccolo F, Halawani AR (2014) The concept of exception: from politics to spatial domain. Planum 2(29):1–11

Lo Piccolo F, Todaro V (2015) Latent conflicts and planning ethical challenges in the South-Eastern Sicily "Landscape of exception". In: Milan Macoun KM (ed) Book of Proceedings Annual AESOP 2015 Congress Definite Space - Fuzzy Responsibility. České vysoké učení technické v Praze, Prague, pp 2534–2544

Lo Piccolo F, Todaro V (2018) L'invisibilità sociale degli immigrati nella Sicilia post-rurale: il caso della "fascia trasformata" del ragusano. In: Lo Piccolo F, Picone M, Todaro V (eds) Transizioni postmetropolitane. Declinazioni locali delle dinamiche posturbane in Sicilia. FrancoAngeli, Milan, pp 285–306

Lo Piccolo F, Picone M, Todaro V (2017a) South-eastern Sicily: a counterfactual post-metropolis. In: Balducci A, Fedeli V, Curci F (eds) Post-metropolitan territories. Looking for a New Urbanity. Routledge, Abingdon, pp 183–204

Lo Piccolo F, Picone M, Todaro V (2017b) La Sicilia sud-orientale, una regione post-metropolitana controfattuale. In: Balducci A, Fedeli V, Curci F (eds) Oltre la metropoli. L'urbanizzazione regionale in Italia. Guerini e Associati, Milan, pp 223–250

Martin P (1985) Migrant labor in agriculture: An international comparison. Int Migrat Rev 19(1):135–143

Martin P (2002) Mexican workers and U.S. agriculture: The revolving door. International Migration Review 36(4):1124–1142

Medici Senza Frontiere (2007) Una stagione all'inferno. Rapporto sulle condizioni degli immigrati impiegati in agricoltura nelle regioni del Sud d'Italia. Medici Senza Frontiere Onlus, Rome

Mendoza C (2001) Cultural dimensions of African immigrants in Iberian labour markets: a comparative approach. In: King R (ed) The Mediterranean Passage. Migration and New Cultural Encounters in Southern Europe. Liverpool University Press, Liverpool, pp 41–65

Mignella Calvosa F (2013) Premessa. In: Colloca C, Corrado A (eds) La globalizzazione delle campagne. Migranti e società rurali nel Sud Italia. FrancoAngeli, Milan, pp 9–12

Ministero del Lavoro e delle Politiche sociali (2012) Secondo Rapporto annuale sul mercato del lavoro degli immigrati, Rome. http://www.lavoro.gov.it/Priorita/Documents/II_Rapporto_imm igrati_2012.pdf

Osti G (2010) Fenomeni migratori nelle campagne italiane. Agriregionieuropa 22(6):59–60

Pugliese E (2006) L'Italia tra migrazioni internazionali e migrazioni interne. il Mulino, Bologna

Reyneri E (2007) La vulnerabilità degli immigrati. In: Brandolini A, Saraceno C (eds) Povertà e benessere. Una geografia delle disuguaglianze in Italia. il Mulino, Bologna, pp 197–236

Scucces Ferraro RR (2006) Studio osservazionale sulla IVG nella Ausl 7 di Ragusa negli anni 2004–2005. Sanità Iblea V(4):6

Todaro V (2014) Immigrati in contesti fragili, tra conflitti latenti e limiti delle politiche locali di accoglienza. Urbanistica Informazioni 257:42–45

Todaro V (2015) La "pianificazione" del paesaggio come strumento di controllo sociale. Gli immigrati nelle serre del ragusano, tra produzioni di qualità e negazione dei diritti di cittadinanza. In: Atti della XVIII Conferenza Siu, Italia 45–45. Radici, condizioni, prospettive, 11–13 June 2015, Venice. Planum Publisher, Rome-Milan, pp 927–932

Todaro V (2016) Transizioni post-metropolitane ai margini: la Sicilia dei migranti, oltre l'invisibile. Territorio 76:72–77

Todaro V (2017) Mutamenti spaziali come effetto di mutazioni sociali? Questioni aperte sui flussi migratori nei territori dell'agricoltura di qualità in Sicilia. Contesti 1–2:72–87

Vertovec S (2007) Super-diversity and its implications. Ethnic and Rural Studies 30(6):1024–1054

Zanfrini L (1993) Gli immigrati nei mercati del lavoro locali. Spunti di riflessione dalla ricerca empirica. In: Colasanto M, Ambrosini M (eds) L'integrazione invisibile. L'immigrazione in Italia tra cittadinanza economica e marginalità sociale. Vita e Pensiero, Milan, pp 33–112

The Power of Fiction in Creating a Territory's Image

Vincenzo Todaro, Annalisa Giampino, and Francesco Lo Piccolo

Abstract The current economic crisis, and its effects on the European sovereign debt, has resulted in a significant reduction in Italian investments in the field of preservation and enhancement of cultural heritage. In fragile contexts, such as Sicily, the contraction of investments has put further into crisis the already weak management and planning system. Despite this scenario of crisis, in South-Eastern Sicily, the province of Ragusa shows a significant resilience to the crisis also due to the "Montalbano effect" and the capability to rethink the traditional heritage policies (Magazzino and Mantovani 2012). The paper, based on the analysis of the South-Eastern Sicily case study, reflects on these potential and conflictive elements generated by the relationship between cine-tourism and local policies.

1 Media Image in the Creation of the New Urban Dimension

During the transition to postmodernity, the role of media images has been found to be decisive in the narration and consequent *definition* of the post-metropolis. In particular, according to Shiel (2001) cinema, as stories in images, has been one of the greatest forces in globalisation processes in the twentieth century, making a significant contribution to the construction of the post-metropolis image.

However, the synthesis process that cinematographic image produces compared to complexity of real life can at times distort it; or it can produce types of "space

V. Todaro (✉) · A. Giampino · F. Lo Piccolo
University of Palermo, Palermo, Italy
e-mail: vincenzo.todaro@unipa.it

A. Giampino
e-mail: annalisa.giampino@unipa.it

F. Lo Piccolo
e-mail: francesco.lopiccolo@unipa.it

F. Lo Piccolo et al. (eds.), *Urban Regionalisation Processes*,
UNIPA Springer Series,
https://doi.org/10.1007/978-3-030-64469-7_8

165

consumption" (Connell 2012) as it alters the authenticity of places that we are not always aware of or able to control. Soja (2000) seems to be aware of this, highlighting how the relationship between image and reality introduces the concept of "hyper-reality", that highlights the alterity of a new dimension, both the first and the second.

Therefore, we should ask how the image of certain places transposes reality into a new urban dimension and what type of imaginary relationship it can maintain with reality. Also, what the actual spatial repercussions are of the relationship between image and reality and how much of these places and people are aware of.

To consider these matters further, in reference to South-Eastern Sicily, this Chapter wishes to look at the media image produced by the TV fictional series "Inspector Montalbano". By moving from the analysis of the effect on tourist flows and of spatial effects caused in the area, the considerations below address a scenario of general economic crisis, in a fragile, marginal territorial context, which tends to find opportunities for expected (tourist) development in the fleeting medium of cinema that is controlled by other parties.

2   Economic Crisis and Cultural Heritage in South-Eastern Sicily

The effects of the 2008 economic crisis caused a clear reduction in state investments in safeguarding and optimising cultural heritage and activities, resulting in a decrease in investment from 0.3% of the GDP in 2009 to 0.17% in 2012.

Although signs of recovery have been noted recently (2015), with an investment in cultural services of 0.36% of the GDP, this value is still one of the lowest in Europe (European average 0.45%) and is about half of France's investment (0.73%) (ISTAT 2017). This condition has to deal in Italy with the extent and territorial distribution of its cultural heritage, that requires protection, economic development, optimisation and promotion all at the same time.

Also, the notable reduction in state spending on cultural heritage management was accompanied by a partial reduction in local body spending, although the latter had to compensate for the reduction in state investments.

As found in the 2017 BES Report (ISTAT 2017), local councils autonomously contributed about two-thirds of public spending for cultural activities. As can be seen in Table 1, from 2009 to 2014 spending in the cultural activities and environmental heritage protection and optimisation sector decreased by 59.16%.

Also, although Sicily has a large concentration of cultural heritage compared to other Italian regions, as maintained by Casavola and Trigilia (2012), it does not show itself to be effective in optimising such a resource. And it is also in light of the distance between cultural heritage *ownership* and activation of the optimisation policies, that, in political rhetoric but also in real-life economy, the gap between Northern and Southern Italy increases (Lombardo 2016).

Despite this scenario of crisis, South-Eastern Sicily is an exception, as it maintains its tourists, also thanks to the so-called "Montalbano effect" (Magazzino and Mantovani 2012).

Table 1 Public expenditure in Sicily in the field of protection and enhancement of cultural heritage

Year	Organisation of culture and related structures (Euros)	Nature protection, environmental assets, parks and reserves (Euros)	Total expenditure in the cultural heritage sector (Euros)
2009	372,630,684	138,180,789	510,811,473
2010	250,708,966	44,307,393	295,016,359
2011	279,559,964	85,219,140	364,799,104
2012	168,238,513	43,739,873	211,830,894
2013	190,027,455	47,592,381	237,619,813
2014	178,673,509	29,891,577	208,565,086
2015	257,969,932	41,449,355	299,419,287

Ragioneria Generale dello Stato, Bilanci delle Regioni e delle Province Autonome, years 2009–2015

The widespread presence of cultural heritage has formed a historically stratified landscape over the years, which has a great identity value. This has made South-Eastern Sicily an ideal location for several, prestigious cinema productions, including the TV fictional series "Inspector Montalbano", produced by Palomar for the National Italian Television (*Radio Televisione Italiana, RAI*).

The characteristic of this territory lies in the stratification of heritage and resources of different kinds, that depend highly on each other: sand dunes on the coast, Greek-Roman archaeological sites, agricultural landscapes filled with country houses and Arabic farms, medieval old towns and Baroque cities. All this defines a cultural landscape that provides *natural* scenic settings for cinema filming. These environments are places that accompany the usual reflections by Inspector Montalbano and that Roskill (1997, 202) defined as "landscapes of presence", as they are the most suitable setting for the film (Fig. 1).

Moreover, the other part of the cultural heritage that is the backdrop for the TV series is the extraordinary urban landscape that exploits the natural conformation of the ground and the theatrical nature of the urban wings, with their Baroque architecture, making for ideal scenographic settings.

The façades of palaces and churches, accentuated by wide, deep staircases, and their shiny domes, show themselves off to the surrounding landscape as unique scenic settings, that dominate the surrounding small, low buildings (Fig. 2).

More generally, these are part of a network of small- and medium-sized urban areas, an expression of urban planning and Baroque (Noto, Scicli, Rosolini, Modica, Ragusa, Ispica) (Figs. 3–4) and Liberty architecture (Ispica, Canicattini Bagni).

The "fictional towns" of "Montelusa" and "Vigata" in the TV series show Val di Noto as a single entity in the international Baroque panorama, characterised by the contemporary reconstruction of several towns (Caltagirone, Catania, Militello Val Catania, Modica, Noto, Palazzolo Acreide, Ragusa, Scicli) after the terrible earthquake in 1693 when 93,000 people died and 60 towns were seriously damaged.

Fig. 1 The "Fornace Penna" along the coast of Scicli (Photo by Giuseppe Abbate)

Fig. 2 San Giorgio's Cathedral in Modica (Photo by Giuseppe Abbate)

Fig. 3 The centre of Scicli (Photo by Giuseppe Abbate)

Fig. 4 The centre of Ragusa (Photo by Vincenzo Todaro)

3 Cultural Heritage Development Policies in South-Eastern Sicily[1]

There are two factors that have brought South-Eastern Sicily to optimise its cultural heritage and relaunch its own cultural and tourist image. The first is connected to the recovery of Ragusa Ibla historical old town, that was started by means of RL 61/1981. The actions taken for urban and building upgrading and recovery have brought about the opening of several accommodation facilities (Lo Piccolo et al. 2015).

The second factor, promoted by the Syracuse, Catania and Ragusa Cultural Heritage Departments, which carry out the task of protecting the area's cultural heritage, is seen in the *construction* of the joint image of the "late-Baroque cities". This territorial image, built around recognition of the urban and Baroque architectural culture as a unifying identity value, is based on the cultural enhancement policies started in the period 1990–2000.

One of the outcomes of these policies is the inclusion of "late-Baroque cities of Val di Noto" (including the cities of Noto, Scicli, Ragusa, Militello Val di Catania, Caltagirone, Palazzolo Acreide, Catania, Modica) in the "UNESCO World Heritage List" in 2002. Regional funding was also provided for the establishment of *tourist districts* aimed at promoting the "Baroque of Val di Noto" also for the high-quality food and wine productions.

What effect to these cultural enhancement policies have on tourism? What are the repercussions on the improvement in the tourist offer?

Actually, in the last twenty years, South-Eastern Sicily has thoroughly transformed its territorial profile. There has been a dual effect: differentiation of the types of available accommodation, from large hotels and tourist holiday villages to the creation of small and medium-sized hotels; and their distribution throughout the area, compared to the concentration on the coastline that was the case in the past. In particular, the type of accommodation that has shown a significant increase in the last ten years are not hotels (farmhouse accommodation and B&Bs).

With regard to tourist flows, some sector studies (Mantovani 2010; Magazzino and Mantovani 2012), that cross-check different databases (tourist density and tourist specialisation), clearly show how the arrivals and presences in the province of Ragusa—compared to the whole of Sicily—in the period 2000–2008, show growth of about 5.00% and 5.80%, compared to the same indicators (4.50% and 5.20%) in the previous decade (1990–1999). Also, in the subsequent period (between 2009 and 2010 and between 2012 and 2013), compared with the wider panorama of the international crisis that also affected the tourist sector, South-Eastern Sicily generally seemed to maintain positive figures.

When speaking of nationality, the presence of Italians is much more widespread than that of foreigners, which up to 2009 stood at about one third of the former; however, this difference started to decrease starting from 2010: in 2013, there were 72,354 foreign individuals present, compared to 128,555 Italians.

---

[1] Data and considerations of this section do not take into account the effects connected to the Covid-19 crisis.

Some of the most prominent foreign nationalities in 2013 were French (22,667), German (8, 523), American (5169), British (4950), and Swiss (4738).

Moreover, compared to the above scenario, although it is not possible to link the reasons for the holding of tourist flows solely to "Inspector Montalbano", it is agreed that this TV series has played a decisive role in the international promotion of South-Eastern Sicily's image.

There have been countless presences in recent years, in fact, around the so-called "Montalbano locations". The transfer of Scicli town hall, the location used as the "Vigata" police station in the series, is the consequence of the high numbers of tourists interested in visiting "Inspector Montalbano's office" (Fig. 5).

The inspector's house, located on the beach at Punta Secca (Santa Croce Camerina), has become the "La casa di Montalbano" B&B (Figs. 6–7). The *RAI* TV crew has to make a *group* booking to find availability for the rooms to record the new episodes of the TV series.

With regard to this phenomenon, and some clear distortions that this brings with it, it must however be pointed out that for the Ragusa area tourism is a recently-formed opportunity, that needs more structured strategies and greater consolidation (Azzolina et al. 2012).

The same experience of several tourist districts, that are created in this territory through cooperative spirit, has been a mere aggregation of municipalities that cannot express a "joint vision of South-East territory" (Azzolina et al. 2012, 161). The current tourist offer, rather than responding to the new system logic, continues to propose out-of-date, traditional models aimed at developing hotels and catering, neglecting facilities and services.

This model appears to be inadequate for the current time, as it does not respond to the demand for a differentiated tourist offer that is integrated with infrastructures and services (including technological ones) to help make use of cultural heritage, a realm in which it is possible to effectively compete in international tourist markets.

4 The "Montalbano Effect"

The TV fictional series "Inspector Montalbano", adapted from the novels by Andrea Camilleri, initially created an increase in tourist flows in South-Eastern Sicily in the period 1999–2007 and a limitation of recession trends in the period 2007–2013 compared to the regional and national context.

As stated by Mantovani (2010), based on an analysis of tourist development indicators, the "Montalbano effect" has permitted South-Eastern Sicily and Ragusa in particular to limit the drop in tourist numbers in the period of international financial crisis. However, the "Montalbano effect" has not only driven tourist development; it has also been an opportunity to rethink local cultural heritage enhancement policies, via the promotion of cinema-tourism.

High profile tourist destinations increasingly sponsor their own territory with marketing operations that use film sets and scenes shot in those places as leverage.

Fig. 5 The *police station* at the Town Hall of Scicli (Photo by Giuseppe Abbate)

Fig. 6 Inspector Montalbano's house (Photo by Vincenzo Todaro)

Fig. 7 Inspector Montalbano's house and the coastal urbanisation of Punta Secca, Santa Croce Camerina (Ragusa) (https://www.google.it/maps/)

However, as pointed out by Joanne Connell (2012), cinema tourism probably has a historical precedent in literary tourism (Pocock 1992; Squire 1994; Busby and Hambly 2000; Robinson and Andersen 2002; Hoppen et al. 2014), although there are significant differences: the images produced via reading require the reader's imagination to be created (Butler 1990; Verdaasdonk 1991); consequently, they tend to produce different, higher tourist expectations on a given destination, as they are not pre-established as cinema images are (Pocock 1992). On the relationship between literary and cinema tourism, the Di Betta study (2015) applied to the case of "Inspector Montalbano" compares the different places that refer to each of the two categories: Agrigento for literary tourism and Ragusa for cinema tourism.

The study results show how literary stories are no less effective than television in terms of effects on tourist flows; to the contrary, the two areas seem to be complementary to system policies that acknowledge the added value of integrated tourism.

However, Joanne Connell (2012) believes that in the ambiguous relationship between reality and cinema image, it is possible to distinguish at least two different approaches: the construction of the film image takes place with the clear intention of maintaining the authenticity of reality as much as possible; or through specific manipulation of reality with cinema's main tools, with editing being the main one.

In the latter case, there is a more controversial and sometimes contradictory change in the perception of real space by film-making; a kind of a real *consumption of space* which occurs by removing or adding other elements and that produces an incredible paradox: the boundary between reality and film image becomes indistinguishable (Schofield 1996); the film image takes over from reality, staying in the memory of those who have witnessed it; in visiting places, the observer records an important difference between image and reality, to the extent that the latter is not recognised as such; this is the cause of frustration in the tourist experience (Connell and Meyer 2009).

The film image generates a spatial *appropriation* of the place and consequently the reality loses respect, as it is no longer recognisable. Two examples show the spatial effects that this phenomenon creates.

The first concerns the case of "Inspector Montalbano": in the 1990s, the town of Porto Empedocle (birthplace of A. Camilleri) asks for and obtains permission from the writer that the town can add "Vigata" to its own name, to honour (with a clear return on tourist image) both him and his stories (Fig. 8). In this case, the film image appropriates the real space.

To the contrary, the second concerns the phenomenon of loss of recognisability of places compared to the predominance of the film image over reality: this is the case reported by Connell (2012), of an American producer who, while looking for a film location in Scotland in the 1950s, returned to the United States after not finding anywhere that *looked like* Scotland.

Fig. 8 Identification of the city of Porto Empedocle as "Vigata" (http://jucaffe.blogspot.it/2013/05/la-forma-dellacqua-camilleri-speciale.html)

5 Cinema Tourism and Cultural Heritage Enhancement Policies in Sicily

The successful television series, produced by Palomar for the *RAI*, and also thanks to its broadcasting on *RAI International* in 18 different countries[2] has been a driving force for international tourism in a peripheral area of Sicily, namely the South-East. Ragusa, Scicli, Santa Croce Camerina—some of the locations used to film the series—have become destinations for tourists who want to see the places around which the fictional series is centred.

The success of the series and the consequential benefits for the area were recognised in 2007. In fact, through RL 16/2007, the Sicilian Regional Administration established a "Regional fund for cinema and audio-visual works" for the production of films to be made in Sicily, called the "Film Commission Regione Siciliana" to manage said funds.

---

[2]The TV series has been broadcast in Argentina, Australia, Austria, Bulgaria, Czech Republic, Denmark, Finland, France, Germany, Hungary, Lithuania, Poland, Romania, Russia, Spain, Sweden, United Kingdom and USA.

In reference to the above, two new elements emerge, although highly problematic compared to traditional heritage enhancement policies.

An initial aspect concerns recognition of the media image when promoting cultural heritage. It is no coincidence that the law states that some scenes must be filmed in pre-decided places in order to promote and spread the image of Sicily more effectively.

However, this recognition is not without its ambiguous side. In fact, the race to offer locations gave rise to a high rate of competition between municipal councils, to the detriment of an overall enhancement policy for the area's cultural system.

Following the economic recession, and the simultaneous reduction in municipal finances, the latter realised the need to set up partnerships to promote their own cultural heritage through film productions. Therefore, on 11 February 2015, towns in the province of Ragusa[3] signed a "Protocol of Understanding for the coordination and planning 2015–2017 of activities to support the production of Inspector Montalbano", also setting out intervention modes and joint actions.

While the use of media image has brought about this change in standard in the public subject's action model, with regard to cultural heritage enhancement, on the other hand, there is an ambiguity in the power of the image to convey an ideal reality that does not correspond to the actual reality.

The unspoilt coastal area, with its sand dunes, as seen in the TV series, does not, in fact, reflect the real situation on the coastlines of South-Eastern Sicily. Illegal buildings, the man-made load linked to holiday homes and hotels, and the presence of production infrastructures and structures have all transformed the coastal areas in the province of Ragusa into a highly-affected, fragile habitat.

According to a study carried out about the state of urbanisation in Sicily, in support of the regional urban plan, 80.5% of the urbanised coastline in the Ragusa area within 150 metres from the shore is illegal; and 88.6% of the buildings located in the area 150–500 metres from the shoreline are illegal (Trombino 2005).[4] This data shows the conflictual nature of the Sicilian situation, where cultural heritage protection and enhancement policies and urban-planning policies seem to be moving in completely separate directions.

The second aspect concerns the possibility of creating public-private partnerships for the promotion of cultural heritage, with a clear reduction in public investment in this sector.

In return for the employment of local human resources (actors, interns, master tradesmen, etc.) and a pre-set percentage of filming of Sicilian territory and its heritage (20% of the total of outdoor filming of the edited film and at least 30% of the total footage of the edited film), RL 16/2007 provides a non-repayable contribution to film productions.[5]

---

[3]The town councils which signed the protocol are: Ragusa, Acate, Chiaromonte Gulfi, Comiso, Giarratana, Ispica, Modica, Monterosso Almo, Pozzallo, Santa Croce Camerina, Scicli and Vittoria.

[4]RL 78/1976 states a restriction on all constructions within 150 metres of the shoreline.

[5]Funding contributions are issued based on the number of weeks of filming in Sicily. They range from three annual contributions, from an overall maximum of 250,000 Euros if filming goes beyond seven weeks, to a minimum of 10,000 Euros for filming of less than one week.

For the promotion of cultural heritage, this is a form of joint-funding for territorial marketing, entirely the responsibility of the public entity. This public-private partnership is repeated again in the 2015 "Protocol of Understanding" which, in return for the municipal council's undertaking, sees the company Palomar involved in transforming public assets provided by the councils into permanent exhibition spaces that will be managed as a partnership.

It is clear that in times of economic crisis, film productions are an effective means for the enhancement of cultural heritage, as can be seen in the "Montalbano effect" being the driving force for Ragusa heritage enhancement by the public entities involved.

However, although this change in standard can provide a substantial change to the *enhancement* of the territory during a permanent economic crisis period, it also raises several doubts of an ethical nature regarding the use of public assets for private purposes, even within partnerships. The Sicilian Regional Administration does not seem to have noticed this, however, as when faced with the possibility of transferring the location for future filming of the TV series to Apulia, it approved a bill for "Interventions in favour of audio-visual serial productions of a continuous and cyclical nature", known as "Montalbano" *DdL* ("Montalbano" Draft Law), that provides for 50,000 Euros funding for each episode of at least 45 minutes, for film productions and TV series with a history of at least three years of programming and production, and a total allocation of spending of 2,000,000 Euros.

6   Conclusions

Several authors have studied the effects of film and TV productions on tourism. These analyses have mainly focused on the influence of the locations' image on the choices of sets and repercussions on tourist flows (Schofield 1996; Hudson and Ritchie 2006a, b; Soliman, 2011).

However, the effects of film productions on the planning and enhancement policies for the locations' cultural heritage has not really been considered by the literature on this subject. The case study of South-Eastern Sicily shows how cinema tourism has had considerable effects on the territory, on cultural heritage and on the parties involved in managing them.

By moving the point of view from the users of the locations to the cultural heritage that acts as a setting for the television and film productions, the matters of planning and managing cultural heritage, as the contribution proves, are an especially important field of investigation.

The results obtained show how the so-called "Montalbano effect" has had at least two positive impacts in a period of crisis for public funding:

- The maintaining of the tourist industry linked to the use of the Hyblaean area cultural heritage, even during a recession;

- A change in standard in cultural heritage management and enhancement models, creating both greater cooperation between the public entities involved and between the public entities and the private players, in order to optimise available resources.

With regard to the first point, the province of Ragusa has shown considerable resilience to the crisis that also affects the tourist sector, over time maintaining the number of foreign presences and recording a constant and significant increase in accommodation available other than classic hotels (farmhouse accommodation and B&Bs). There appears to be only a weak, or even non-existent system strategy, however, to direct tourist flows and demand towards a more sustainable, responsible and innovative type of tourism. The current offer is based on a short, fragmented, weak and uncompetitive chain compared to the innovative content and services linked to the offer of cultural assets.

The second point, regarding the economic synergy between public and private funding, as Soliman (2011) suggests in relation to the increase in the types of film tourism, could introduce forms of free taxation and special incentives for film productions rather than resorting to *emergency* laws to favour productions, as used in the Ragusa area.

What is described above represents the initial trials of innovation in the cultural heritage sector, which, as stated by Connell and Meyer (2009), in order to be effective must not cause any cultural or environmental damage to the heritage, but rather have a greater enhancement effect on them. The data found shows the parties' attention to types of historical and architectural heritage protection, on the one hand, and on the other a lack of attention for the territories and landscape that are not included in the "Protocol of Understanding" signed by the town councils in the Hyblaean area. At the same time, urban planning policies do not often interact with territorial and environmental asset protection and enhancement policies.

In light of the considerations made, rather than referring to a condition of "hyperreality" (Baudrillard 1983; Soja 2000), owing to the spatial repercussions that it has on the area, the relationship between image and reality created by the TV series "Inspector Montalbano" on South-Eastern Sicily is more a case of evoking a *postreality*, where the parts of reality tend to become more tenuous than the imaginary ones which, to the contrary, seems to be increasingly real.

References

Azzolina L, Biagiotti A, Colloca C, Giambalvo M, Giunta R, Lucido S, Manzo C, Rizza S (2012) I beni culturali e ambientali. Ragusa. In: Casavola P, Trigilia C (eds) La nuova occasione. Città e valorizzazione delle risorse locali. Donzelli, Rome, pp 151–162

Baudrillard J (1983) The precession of simulacra. In: Baudrillard J (ed) Simulations. Semiotext(e), New York, pp 1–79

Busby G, Hambly Z (2000) Literary tourism and the Daphne Du Maurier festival. In: Payton P (ed) Cornish studies. Exeter University Press, Exeter, pp 197–212

Butler R (1990) The influence of the media in shaping international tourist patterns. Tourism Recreation Research 15:46–53

Casavola P, Trigilia C (eds) (2012) La Nuova Occasione. Città e valorizzazione delle risorse locali. Donzelli, Rome

Connell J (2012) Film tourism e Evolution, progress and prospects. Tour Manag 33(5):1007–1029

Connell J, Meyer D (2009) Balamory revisited: an evaluation of the screen tourism destination-tourist nexus. Tour Manag 30(2):194–207

Di Betta P (2015) L'effetto di Montalbano sui flussi turistici nei luoghi letterari e televisivi. Economia e diritto del territorio 2:269–290

Hoppen A, Brown L, Fyall A (2014) Literary tourism: opportunities and challenges for the marketing and branding of destinations? Journal of Destination Marketing & Management 3(1):37–47

Hudson S, Ritchie B (2006a) Film tourism and destination marketing: the case of Captain Corelli's Mandolin. Journal of Vacation Marketing 12(3):256–268

Hudson S, Ritchie B (2006b) Promoting destinations via film tourism: an empirical identification of supporting marketing initiatives. Journal of Travel Research 44(2):387–396

ISTAT (2017) BES 2017. Il benessere equo e sostenibile in Italia. ISTAT, Rome

Lo Piccolo F, Giampino A, Todaro V (2015) The power of fiction in times of crisis: movie-tourism and heritage planning in Montalbano's places. In: Gospodini A (ed) Proceedings of the International Conference on Changing Cities II: Spatial, Design, Landscape Socio economic Dimensions. Porto Heli, Peloponnese, Greece, 22–26 June 2015, Grafima Publ, Thessaloniki, pp 283–292

Lombardo U (2016) Turismo culturale e crescita economica: un'analisi sui rendimenti e i livelli di attivazione dei beni culturali in Sicilia. StrumentiRes 2:1–24

Magazzino M, Mantovani M (2012) L'impatto delle produzioni cinematografiche sul turismo. Il caso de «il Commissario Montalbano» per la Provincia di Ragusa. Rivista di Scienze del Turismo 1:29–42

Mantovani M (2010) Produzioni cinematografiche e turismo: le politiche pubbliche per la localizzazione cinematografica. Rivista di Scienze del Turismo 3:81–103

Pocock D (1992) Catherine Cookson country: tourist expectation and experience. Geography 77:236–243

Robinson M, Andersen HC (2002) Literature and tourism. Essays in the reading and writing of tourism. Continuum, London

Roskill M (1997) The language of landscape. Pennsylvania University Press, Philadelphia

Schofield P (1996) Cinematographic images of a city: alternative heritage tourism in Manchester. Tour Manag 17(5):333–340

Shiel M (2001) Cinema and the city in history and theory. In: Shiel M, Fitzmaurice T (eds) Cinema and the city: film and urban societies in an urban context. Blackwell, Oxford, pp 1–18

Soja EW (2000) Postmetropolis: critical studies of cities and regions. Blackwell, Malden (MA)

Soliman DM (2011) Exploring the role of film in promoting domestic tourism: a case study of Al Fayoum. Egypt. Journal of Vacation Marketing 17(3):225–235

Squire S (1994) Gender and tourist experiences: assessing women's shared meanings for Beatrix Potter. Leisure Studies 13:195–209

Trombino G (2005) Le coste: urbanizzazione ed abusivismo, sviluppo sostenibile e condoni edilizi. In: Savino M (ed) Pianificazione alla prova nel Mezzogiorno. FrancoAngeli, Milan, pp 279–291

Verdaasdonk D (1991) Feature films based on literary works: are they incentives to reading? On the lack of interference between seeing films and reading books. Poetics 2:405–420

Institutional Policies

In Search of the Metropolitan Dimension

Francesca Lotta

Abstract The new settlement phenomena in our territory, linked to current globalisation processes and the increasing complexity of contemporary society, have
caused a crisis of the historical boundaries and are creating new planning layouts. In
this context, the metropolitan dimension is once again viable, and Italy, with Sicily
sometimes experiencing these processes in advance, is reorganising the institutions
covering large areas, directing processes and trying out new convergences, under the
driving force of new public finance reorganisation policies.

1 New Settlement Layouts in Italy

In the 1960s, the national territory was involved in the new debate about territorial
layout. In the industrial triangle of the time, the need had arisen to define and identify
that territory using a different logic than the historical provincial borders. But at that
time, the idea was not realised, and it was only the discussions of almost a decade
later[1] that addressed the metropolitan issue again, tangibly.

---

[1] In 1960, the Italian government started a path of modernisation for the country. The results of this
debate were gathered together under the title *Progetto 80.*

F. Lotta (✉)
University of Palermo, Palermo, Italy
e-mail: francesca.lotta@unipa.it

After the initial (and actually the last) debates which lasted about twenty years,[2] Italy tried once again in the early 1990s to navigate in the huge dimension, following a path laid out by Europe. Driven by the need to define new boundaries at that time, the law-making body created an overlap of authorities, creating problematic situations to manage in a large part of the country.

In the last decade, new financial needs have driven the national government to return to the complex matter of reorganising territory, managing processes and experimenting with new convergences. For this reason, the *reorganisation* of large area bodies and the establishment of metropolitan cities have begun once more. Today, in fact, it is possible to photograph a new layout and we can explore local attempts to provide quick answers to the national government, which are not without inconsistencies.

2 Administrative Re-organisation of the Vast Areas in Sicily

As has often occurred in the past, Sicily, compared to the general Italian context, is a precursor of the metropolitan issue. Thanks to its autonomy over several areas and in concurrence with the establishment of regional provinces, by means of RL 9/1986, Sicily regulated the criteria for identifying and outlining its metropolitan areas: conglomerates of several urban centres. The law at that time defined the functions and goals to be achieved (article 21) and resolved the matter of the overlapping of territorial authorities. In fact, it placed the metropolitan areas under the provincial bodies. The regional identification criteria of the areas were mainly quantitative and referred to a monocentric model.

In detail, the resident population of the entire area had to be greater than 250,000 inhabitants and grouped around a municipality of at least 200,000 inhabitants. Alongside this vision focused on the metropolitan area, there was a marginal reflection about the importance of essential services, transport system and the economic and social development of the entire area (article 19).

The Regional Presidential Decree (*Decreto del Presidente della Regione, DR*) to establish metropolitan areas was signed by the Region's President on 10 August 1995, after the regional implementation of NL 142/1990, that sped up the issue of the decree of establishment.

---

[2]From the end of the 1950s to the 1970s, Italy and above all the cities in the North, which could attract new residents and a large number of commuters, went through a period full of cultural ferment. In 1958, the attentive geographer Aldo Sestini paid testimony to the rapid, radical change in society. Those years saw the first theoretical efforts to identify scientifically valid criteria for defining metropolitan areas. Sestini (1958) and Aquarone (1961) are remembered for their most interesting considerations for the identification of metropolitan areas. The former believes the percentage of territorial population density and that of non-agricultural profits were fundamental. The latter considered the industrial and service characteristics and the size and frequency of trade and relations between the central area and the surrounding areas.

The only territories that corresponded to these administrative and dimensional system criteria were the Metropolitan Areas of Palermo, Catania and Messina. These three new areas made no reference, however to the economic characteristics, or to the spatial, environmental, infrastructural or even settlement implications (Di Leo 1997; Schilleci 2008).

In Sicily, like in Italy, the above identification had no operational follow up: the identified bodies were only empty containers, without an organic structure that could fulfill the functions set out by law. Once again, the metropolitan dimension was forgotten about.

After almost twenty years, under national influence, the regional government returned to the matter of territorial reform and began the procedure for drawing up a bill (resolutions 313 and 354) aimed at establishing metropolitan cities.

With RL 8/2014—entitled "Establishment of free municipal consortia and metropolitan cities"—3 Metropolitan Cities were actually established: Palermo, Catania and Messina.

Once again, the identification criteria were the ones contained in the previous law, that suggested the monocentric model: territorial continuity and number of inhabitants of at least 250,000 for the metropolitan cities, and 180,000 for the free municipal consortia that were to replace the provinces.

The Sicilian law, unlike the NL, known by the name of the signatory minister, Delrio, introduced the possibility for each municipality to join an adjacent metropolitan city, a free consortium or even establish new ones, by referendum. There were several local debates arising from the recognition of the local Sicilian autonomies, as well as new issues.

The issues, in fact, concerned the head municipalities of the free consortia of Palermo, Catania and Messina, which were residual to the respective Metropolitan Cities, that would not be able to hold any role as provincial capital/leader of two different intermediate bodies.

The second issue concerned the free consortium of Catania: with the introduction of the Metropolitan City of Catania, the residual free consortium would have been split into two areas, north and south, and the necessary criteria of territorial continuity would not have been met. To complete the entire regional picture, the free consortia of Trapani, Agrigento and Syracuse would have fallen within the boundaries of the former provinces.

The above issues of potential dual roles and territorial non-continuity brought the regional government to take a backward step with the procedure already started. With RL 15/2015, the Sicilian government established 3 metropolitan cities and 9 free municipal consortia, copying the boundaries of the former Provinces of the same name and neutralising public debates and consultations.

The result of several uncertain Sicilian steps regarding the metropolitan issue shows the persistence of an administrative political role that tends to perpetuate a system that centralises power and services (Lo Piccolo et al. 2017) (Fig. 1).

The region's anticipatory moves, compared to the Delrio reform, were cancelled by the contradictory repercussions that the regional law was having in relation to national reform.

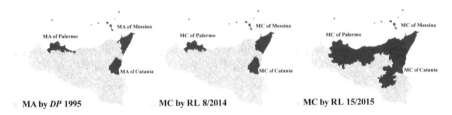

Fig. 1 Configuration of Metropolitan Areas (MA) and Metropolitan Cities (MC) in Sicily, over the years (Image by Francesca Lotta)

The Delrio reform focused on the issues of identity, democracy and plan and the desire to relaunch local systems, to focus new economic and social competitions and innovations (Oliva 2013).

However, Lo Piccolo et al. (2017) maintain that, in Sicily's case, a slow change can also be recognised in the desire to look at planning strategies in the metropolitan context.

The search for a shared future vision, although frustrated by the controversial institutional and administrative decisions, could bring about a democratic public construction process for decision-making and governance processes for the new metropolitan and consortium bodies, together with relative planning tools and deeds.

3   Contemporary Cities and Renewed Boundary Issues

The reorganisation of large area bodies is focused on and develops around the inherent matter of territorial boundaries that are less easily recognisable in contemporary cities (McGrew and Lewis 1992; Nierop 1994). In Italy, the need to identify boundaries, aimed at recognition and governance is part of the concept of territory, as stated in the Einaudi encyclopaedia entry (Roncayolo 1977, 128).

At the same time, the growing complexity of the territory places historical boundaries in crisis, as the result of cultural stratifications that have *adhered* to a given geographical area; in Italy, as in the known American context, there is the question of how to address these boundaries (Soja 2000; Soureli and Youn 2009) and whether they must have an operational role or can be considered as obsolete.

Faced with the new metropolitan dimension, the transformation of two new forms of urban governance could dismantle physical geometries that planning and politics continue to search for, but which perhaps no longer reflect the true territorial relations (Allmendinger and Haughton 2009, 2010; Faludi 2010).

The innovative metropolitan cases on the old continent are multiscale and multi-temporal cases of governance (Bagnasco and Le Galés 2001; Espon 2007; Dematteis and Lanza 2011) where the new spaces for governing contemporary territory have managed to dismantle the historical boundaries.

The Italian approach forces us to think of a physical space defined by limits that often forget about the relations and new balances established in the area. We must, therefore, ask whether the demand for new institutions is false and there is instead a real need for governing territory which has a varying geometry, which, by abandoning a pointless institutional reform, is trying to manage unstable, cross-scale geography, defined according to incidental practices and relational intersections (Amin 2004; Soja 2011; Lieto 2013).

Today, this demand is important in a changing context that has dismantled the urban image that is traditionally linked to the *walled city*, in favour of a new urban layout, without any interruption. No administrative reconsideration aimed at managing these new territories (Deodato 2012) took place for this configuration, and this is due to the rigid pre-existing administrative set-ups, both in terms of aggregation and the maintaining of roles, positions and management tools. This phenomenon is clear, today, in the long identification procedure for the Metropolitan City of Palermo.

4 Identification and Establishment of the Metropolitan City of Palermo

The Regional Presidential Decree (10 agosto 1995), implementing RL 9/1986, identified the Metropolitan Area of Palermo with a territory that included 27 municipalities.[3] The boundaries were defined by quantitative criteria and a monocentric approach: the Metropolitan Area had to have a central hub, and a continuous urbanisation strip along the coast and the two main river valleys, the Eleuterio and the Imera (Schilleci 2008).[4] Even then, however, the borders of the conurbation began to be unclear (Picone 2006). In the time between the above regional law and the issuing of the relative decree, a political-cultural debate sprang up and, in 1988, brought about the proposal of an alternative boundary to the regional government, which was supported by the regional province of Palermo and by *FIAT*-Engineering. The alternative, that remained on paper, was defined by numerical and relational criteria. It included 16 municipalities: 6 representing the actual conurbation and 10 that were close to it (Schilleci 2005, 2008).

The establishment of the Metropolitan Area of Palermo, like in almost all Italian cases, did not then have an operational follow-up. The body remained as an *empty container*, without an organic structure that could fulfill the functions defined by law. Only on 30 January 2014 the Region abolished provincial bodies, anticipating the

---

[3]Palermo, Altavilla Milicia, Altofonte, Bagheria, Balestrate, Belmonte Mezzagno, Bolognetta, Borgetto, Capaci, Carini, Casteldaccia, Cinisi, Ficarazzi, Giardinello, Isola delle Femmine, Misilmeri, Monreale (included in part), Montelepre, Partinico, Santa Flavia, Termini Imerese, Terrasini, Torretta, Trabia, Trappeto, Ustica and Villabate.

[4]The territorial surface area totalled 906 sq.km and a population in 1991 of just over one million inhabitants (1,001,345), amounting to 21.15% of the regional population.

national Delrio provision, and established Palermo Metropolitan City together with Catania and Messina and the 6 free municipal consortia.

The new feature here was the content of articles 2 and 9 where the possibility for each municipality of joining a neighbouring metropolitan city, a free municipal consortium or even establishing new ones, is explained. The geometry resulting from public consultations was different from the ones set out in 1995. In the case of Palermo, in fact, the Municipality of Termini Imerese, moved from the Metropolitan City of Palermo to being a part of the free consortium of Palermo.

However, this new redefinition of boundaries from 2014 has also been amended. Soon after, in fact, the regional government started up the reform to establish metropolitan cities once again and RL 15/2015, using identification criteria contained in the previous legislative provisions, identified the entire former province of Palermo as a metropolitan city.

Once again, the final institutional model has been dictated by purely quantitative criteria that reinforce the monocentric role of the lead city (Fig. 2).

Not even the latest amendments to the RL 15/2015 further to the non-compliance with the NL 56/2014 (so-called "Delrio" Law), and with the Constitution on the definition of free consortia and the direct election of the management bodies for the new associations, managed to change the *new* boundaries and the overall approach to the new territorial dimension. In fact, Palermo had, and continues to want to maintain, a central and centralising function in functional terms and for the concentration of some service categories (Lotta et al. 2017).

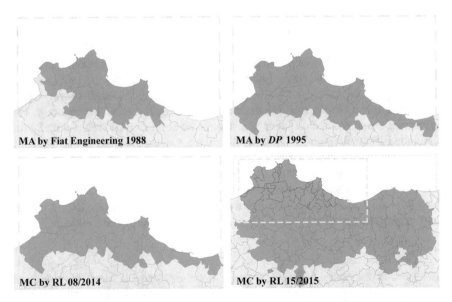

Fig. 2 Configuration of the Metropolitan City of Palermo over the years (Image by Francesca Lotta)

5 Metropolitan Territory Management Prospects

Despite the fact that the territorial boundaries have remained unchanged, the establishment of the metropolitan city has brought with it several academic observations on the management of the territory. These moments have, in fact, proven to be fundamental to discuss the continuation of possible synergies and cooperation on which the current and future well-being of the metropolitan population depends (Gibelli 1999). Synergy and cooperation relations that must necessarily be addressed and planned by territorial government instruments, now changing.

Spatial planning with which the decision-making process for conversion of the use of land and the allocation of user rights is, in fact, a product of the modern state that is unable to interpret and guide a territory in continuous evolution and transformation, helpless before the need to recompose both society and territory (Dematteis 2011).

Its hierarchical approach has, in fact, shown its limits several times, both functionally and economically, but mostly politically. And if we add the sectoral approach to the hierarchical one, we risk enacting a simplified manipulation that obscures the present, if limited interdependence and integration between the required tools on a vast scale. The metropolitan dimension requires relatively stable and legitimate institutions, with clear responsibilities that, to date, are not a part of the reform.

The metropolitan governments cannot seem to be a kind of simple extension of local organisation and the relative institutional problems, but must be the result of political-institutional and economic choices that were outlined by the law-making body based on organisational values and principles, such as cooperation between municipalities, negotiations, partnership, and flexibility of integrated responsibilities.

In contemporary society, planning means starting up processes wherein various players cover the function of points in a high-intensity organisational and decision-making network, that interact through mediation, negotiation and cooperation (Bobbio 2002).

Far from any kind of deregulation model, we are certain that a solution can be found in the concept of governance, which pursues the end of finding territories and effective governing actions, through several types of coordination between players and new, flexible, contractual and participatory partnership approaches.

Examples on large-scale governance can be found in some US metropolitan areas, where the interaction between institutions and an extensive public participation has achieved good results in a highly fragmented context, from an administrative viewpoint. What is even more interesting is the context in north Europe (like in Germany and France) where traditional institutional decentralisation practices in local autonomy, aimed at strengthening territorial classification by central institutions.

These actions, together with connection initiatives between bottom-up operations, guarantee a creation of shared urban and territorial policies, characterised by general interest and procedural models, with the final aim of using the advantages that each of the two models offers and resize the respective limits (Gibelli 1999).

In both Germany and France, there are policies created via central and local cooperation, aimed at indicating strategy, priorities, and suggestions in consistent realms and actions.

The aim is to ensure compatibility of local decisions with national and regional guidelines, not solely for land structure and allocation of use plans, but also for individual, important projects, that respect the political priorities identified by the government.

6 Planning Tools for the Metropolitan City of Palermo

Metropolitan planning actions in the Palermo area can be traced to two moments. The first was started in 1995, in the Palermo area, but was never implemented. This was the study for the General Guidelines for the Intermunicipal Plan for the Metropolitan Area of Palermo submitted in 2001 and which included the general directives on fact-finding analyses; maps and a commercial plan.

The second was the Rough Diagram and Proactive Framework with Strategic Value *(Quadro Propositivo con Valenza Strategica, QPS)* for the Province of Palermo Territorial Plan. In this last tool, whose studies were started in the early 2000s and which to date is not yet approved, it can be deduced how the plan pursues a multi-centric territorial organisation, found in the improvement of transportation systems and in the introduction of new production modes. The intricate relationship between the various tools of a huge area has focused efforts and the lack of a coordination plan has left each municipality to plan its own territory, without any systemic vision.

The greater absence has been and continues to be clear in the approval of some complex programmes. These, intended as contributors of a new kind of project and innovation, have had real repercussions on the territory and, intervening only for the parties, have established unclear relations within a potential huge scheme, with inconsistencies and unexpressed projects confirmed in the relations between municipal planning and complex programmes (Lo Piccolo and Schilleci 2005).[5]

These reflections show how the path to follow to produce a real metropolitan government, with real effects on the territory, is still a long and arduous one. Managing to release regulatory policies, that require marked boundaries when there is no clear objective is a tough task, perhaps more so than observing a reform from close up that has not made any substantial or structural changes aimed at an improved territorial governance.

---

[5]For further information about planning in the Metropolitan City of Palermo, see Lotta et al. (2017).

References

Allmendinger P, Haughton G (2009) Soft spaces, fuzzy boundaries, and metagovernance: the new spatial planning in the Thames Gateway. Environ Plan A 41:617–633

Allmendinger P, Haughton G (2010) Spatial planning, devolution, and new planning spaces. Environ Plan C: Gov Policy 28:803–818

Amin A (2004) Regions unbound: Towards a new politics of place. Geogr Ann 86(B1):33–44

Aquarone A (1961) Grandi città e aree metropolitane in Italia. Problemi amministrativi e prospettive di riforma. Zanichelli, Bologna

Bagnasco A, Le Galés P (2001) Le città nell'Europa contemporanea. Liguori, Naples

Bobbio L (2002) I governi locali nelle democrazie contemporanee. Laterza, Rome-Bari

Dematteis G (2011) Le grandi città italiane. Società e territori da ricomporre. Marsilio, Venice

Dematteis G, Lanza C (2011) Le città nel mondo. Una geografia urbana. Utet, Turin

Deodato C (2012) Le città metropolitane: storia, ordinamento, prospettive. http://www.federalismi. it/nv14/articolo-documento.cfm?artid=21925. Accessed 12 Aug 2019

Di Leo P (1997) Area metropolitana di Palermo. Città e Territorio. Bollettino del Dipartimento della Città e Territorio dell'Università di Palermo 3:72–79

ESPON (2007) Espon Project 2.3.2. Governance of territorial and urban policies from UE to local level. Final report. Espon, Luxembourg

Faludi A (2010) Cohesion, Coherence, Cooperation: European Spatial Planning Coming of Age?. Routledge, London and New York

Gibelli MC (1999) Dal modello gerarchico alla governance: nuovi approcci alla pianificazione e gestione delle aree metropolitane. In: Camagni R, Lombardo S (eds) La città metropolitana: strategie per il governo e la pianificazione. Alinea, Florence, pp 79–101

Lieto L (2013) Disuguaglianze e differenze dello spazio della post-metropoli: temi per un'agenda di ricerca. In: Atti della XVI Conferenza nazionale SIU. Planum 27. http://www.planum.bedita. net/xvii-conferenza-siu-2014-1-pubblicazione-atti-1. Accessed 5 Aug 2019

Lotta F, Giampino A, Picone M, Schilleci F (2017) Sulle tracce della post metropoli: l'area metropolitana di Palermo. In: Balducci A, Fedeli V, Curci F (eds) Oltre la metropoli. L'urbanizzazione regionale in Italia, Guerini e Associati, Milan, pp 193–222

Lo Piccolo F, Schilleci F (2005) local development partnership programmes in sicily: planning cities without plans? Plan Pract Res 20(1):79–87

Lo Piccolo F, Lotta F, Schilleci F (2017) Palermo e la dimensione metropolitana nelle oscillazioni di governance e pianificazione. In: De Luca G, Moccia D (eds) Pianificare le città metropolitane in Italia. Interpretazioni, approcci, prospettive. INU Edizioni, Rome, pp 453–473

McGrew AG, Lewis PG (1992) Global politics: globalization and the nation-state. Polity Press, Cambridge

Nierop T (1994) Systems and regions in global politics: an empirical study of diplomacy. International organizations and trade, 1950–1991. Wiley-Blackwell, Chichester

Oliva F (2013) Città come motore dello sviluppo del Paese. In: Atti del 27th Congress INU, 24–26 October 2013, Salerno

Picone M (2006) Il ciclo di vita urbano in Sicilia. Rivista geografica italiana 113:129–146

Roncayolo M (1977) Territorio. Enciclopedia Einaudi, Einaudi, Turin

Schilleci F (2005) Il contesto normativo in Sicilia. Una difficile pianificazione tra ritardi e resistenze. In: Savino M (ed) Pianificazione alla prova nel mezzogiorno. FrancoAngeli, Milan, pp 189–208

Schilleci F (2008) La dimensione metropolitana in Sicilia: un'occasione mancata? Archivio di Studi Urbani e Regionali 91:147–163

Sestini A (1958) Qualche osservazione geografico-statistica sulle conurbazioni. Studi geografici in onore di R. Biasutti. Supplemento alla Rivista Geografica Italiana 61:313–328

Soja EW (2000) Postmetropolis: critical studies of cities and regions. Blackwell, Malden (MA)

Soja EW (2011) Regional urbanization and the end of the metropolis era. In: Bridge G, Watson S (eds) New companion to the city. Wiley-Blackwell, Chichester, pp 679–689

Soureli K, Youn E (2009) Urban restructuring and the crisis: A symposium with Neil Brenner, John Friedmann, Margit Mayer, Allen J Scott and Edward W Soja. Crit Plan 16:35–58

Urbanisation and Urban Regionalisation Processes in the Metropolitan Area of Palermo

Annalisa Giampino

Abstract This article provides empirical evidence that helps to answer several key questions relating to the regionalisation processes in Southern European cities. Several studies have documented the effects produced by the urban regionalisation process, especially in terms of territorial resource consumptions and the formation of new territorial patterns, overlooking the evaluation of public information that has been lost in terms of space and prerogatives of public action in the new post-metropolitan urban space. Starting from this theoretical framework, the interpretation of the post-metropolitan transition of Palermo through the filter of living in its dual dimension of being the outcome of socio-spatial dynamics and as the main objective of quality of life is an opportunity to reflect on the renewal of planning action tools, especially in the context of institutional absence such as in the South of Italy.

1 Introduction

In 1990 Astengo pointed out the need to investigate the "large structural, social and institutional transformations in the country, examined in its physical aspects that can be assessed both quantitatively and qualitatively" (Astengo 1990, 4) and "current trends, with the prospect of identifying possible corrective actions for the main distortions that are discovered" (*ibidem*). In the last thirty years, in fact, there have been several analytical contributions about urbanisation processes,[1] even if there

A. Giampino (✉)
University of Palermo, Palermo, Italy
e-mail: annalisa.giampino@unipa.it

[1] In the Italian research on the nature of the processes of urbanisation we can recognise two phases of study. The first around the 1960s, characterised by the affirmation and evolution of the monocentric metropolitan model as a result of the *scattered* expansion of large cities. These are the years in which the primacy and pervasiveness of the urban element is understood (Ardigò 1967) and the

is a certain detachment between scientific findings and operational repercussions in the territory.

The quantitative analyses and interpretative images that have been produced have shown important clues to the effects produced by the urban regionalisation process, especially in terms of territorial resource consumptions and the formation of new territorial patterns, overlooking the evaluation of *public* information that has been lost in terms of space and prerogatives in the new post-metropolitan urban space. The extent of urban codes, beyond the conglomerates that are traditionally defined as cities, is not only expressed via spatial dilation and modification of transformation vehicles, but has also affected the field of subjects that now govern the transformation processes and the panorama of lifestyles and aspirations of a society that is continuously losing any homogeneity.

This article intends to document these transformations through an interpretation of what occurred in the territorial realm that the Regional Presidential Decree (*Decreto del Presidente della Regione, DR*) 54 of 1995 identified as the Metropolitan Area of Palermo.[2] A historically hybrid territorial context, where the weak elements that are typical of the global north post-metropolis creation processes overlap with characteristics and phenomena of the south of the world. An *extreme* case, as it has been defined several times (Lo Piccolo 2008; De Leo and Lo Piccolo 2015), that, as Brand and Gaffikin (2007, 284) suggest, "can illuminate the challenges and contradictions involved in a proposition, without laying claim to being typical".

From this point of view, the interpretation of ongoing territorial privatisation processes and the importance of market forces in local choices and living models (Giampino et al. 2014; Giampino 2015) come face to face with examples of informality that range from self-construction, in its endemic form of illegal buildings (Alaimo et al. 1996), to the institutional weakness of institutions, when in extreme cases it does not have a specific collusive system (De Leo and Lo Piccolo 2015).

This is an especially significant case that provides us with interesting points to consider regarding the search for "private happiness" and the rediscovery of "public happiness" (Hirschiman 1982) for post-metropolitan inhabitants, but at the same time confirms how distant politics is from being able to interpret this tension. In the context of this picture, the interpretation of the post-metropolitan transition of Palermo (see Sect. 2), through the filter of living in its dual dimension of being the outcome of socio-spatial dynamics (see Sect. 2.1) and as the main objective of

---

end of the city is recognised as a spatially defined functional unit (De Carlo 1962). A resumption of studies occurs with the approval of the National Law (NL) 142/1990 on metropolitan areas. The debate is structured above all through new toponyms capable of grasping the heterogeneity of forms of low-density urbanisation and the crisis of urban-rural dichotomous readings (Astengo and Nucci 1990; Indovina et al. 1990; Magnaghi 1990; Secchi 1994; Clementi et al. 1996).

[2]The Metropolitan Area of Palermo is made up of a central nucleus, identified with the chief town and its hinterland, and by an urbanised coastal strip comprising 27 municipalities. The municipalities are: Altavilla Milicia, Altofonte, Bagheria, Balestrate, Belmonte Mezzagno, Bolognetta, Borgetto, Capaci, Carini, Casteldaccia, Cinisi, Ficarazzi, Giardinello, Isola delle Femmine, Misilmeri, Monreale, Montelepre, Palermo, Partinico, Santa Flavia, Termini Imerese, Terrasini, Torretta, Trabia, Trappeto, Villabate, Ustica Island.

quality of life that may guide the policies (see Sect. 2.2), is an opportunity to reflect on the renewal of planning action tools and standards, especially in the context of institutional absence, as in the South of Italy (see Sect. 3).

2 The Formation of Post-Metropolitan Palermo Amidst Irregular Urban Regionalisation Processes and Lack of Policies

If the metropolisation processes in the area are linked to "successful territorial formations" (Indovina 2005) that can bring about a broad, differentiated use of the area, in weak contexts like Palermo, recognition of a metropolitan area has historically been seen as a political-administrative necessity rather than as the result of a territorial process. The conglomeration and extension processes of the urban area towards non-urban space (Brenner 2016), in Palermo's case, are linked to real estate yield rather than to the confirmation of a decentralised and territorially restructured economic production model.

Therefore, the post-metropolis of Palermo (Fig. 1) is more significantly the spatial product of *suburbanisation* rather than the result of its *regionalisation.*

The productive function, as a recognised vehicle of the start-up of large area dynamics (Soja 2000; Scott 2008), has been weak and fragmented, not supported by service improvement, logistics and infrastructure policies.

It is no coincidence that after the issue of Regional Law (RL) 9/1986—that established the metropolitan areas in Sicily—the construction industry absorbed one quarter of workers in the local Palermo units (Costantino 2008). The dynamics of the formation of the Metropolitan Area of Palermo do not therefore correspond to a standard, definable post-metropolitan model.

They are more likely the result of different logic and reasoning that have over-lapped in any given territory, essentially as a *neutral* element where incomplete projects and needs can be landed (by the inhabitants), and create possibilities for easy earning (for economic operators and administrations).

Fig. 1 Urban sprawl in the Metropolitan Area of Palermo (Photo by Annalisa Giampino)

However, this interpretation, as supported by Soja (2011), is a superficial and unfair *modern* interpretation of urban space, incapable of taking into account the space in its dynamic, dialectic, conflictual, ideological and socially constructed dimension.

When considering the Palermo area, the socio-spatial perspective was assumed, reinterpreting urbanisation processes in relation to economic, social and policy transformations in the ambiguous, conflictual dialogue between global and local.

2.1 On Urbanisation Processes

The historical and physical identity of the Palermo area has been built on a morphological structure, enclosed in the orographic system (that develops inland) and the coastline, where coastal plains alternate with narrow beaches. A territory where the small agricultural villages have been counterpoints to the seaside villages in a system that has always depended on the City of Palermo (Di Leo 1997; Costantino 2008).

Recent urbanisation processes have kept this functional link with leader cities, on the one hand causing a mostly residential growth of top-tier municipalities and on the other causing an erosion of the coastlines due to the seasonal tourism for the Palermo population (Fig. 2).

Constructing a residential suburban belt is not a new phenomenon. It has occurred continuously in a historical process involving the Palermo area since the 1970s.

Fig. 2 Seasonal complex in the municipality of Carini (Photo by Annalisa Giampino)

In fact, in the 1970s, the completion of the road infrastructures, namely the SS 113, the A29 Trapani-Palermo motorway to the west and the A20 Palermo-Catania motorway to the east and the railway line, took on an imaginary boundary role, beyond the coastline where most man-made constructions were concentrated, aiding the formation of low-density settlement systems which were often illegal, at the start linked to the seasonal exploitation of the coastal areas for tourism.

The entire coastal area of the metropolis has been affected by illegal constructions, often without even the most basic urbanisation work, partly remitted and to a lesser extent demolished (some constructions in Carini, for example) to restore use of the coastal area as public asset.

Exploitation of the territory has not only been carried out via illegal planning, but has also been authorised by Master Plans *(Piani Regolatori Generali, PRG)*, that have provided for large areas of urbanisation in the name of an assumed, sometimes deceptive tourist development. Ambiguous processes where, according to Bianchetti (2003), it is difficult to reconstruct the connections between forms and uses, but also between symbolic, economic and social strategies.

Other phenomena are newer, however, and are a part of that progressive privatisation of space that seems to be linked to the regionalisation of urban space.

Faced with the changes in lifestyle, the split between the city of the rich and the city of the poor (Secchi 2013), and the growing fear of diversity, the Palermo area has seen the formation of new selective and homogeneous communities that exclude themselves in their Privatopias (McKenzie 1994). Fences, walls, surveillance systems, autonomous services show us the existence of a trend towards making living secure in low-density areas, outside the City of Palermo.

As already mentioned earlier (Giampino 2015; Giampino et al. 2017), these are local forms of true gated communities. In many cases, these are residential tourist communities, more recently transformed into permanent homes, as in the case of the residential complexes along the coastline east of the city. In other cases, they are recently-built residential communities to the west, in a period of time from 2000–2008, i.e. before the real estate sector crisis, which intended to absorb a large part of the excess population of the City of Palermo.

In the area in question, 26 complexes similar to gated communities were counted, comprising detached villas that extend from the outlying suburbs of Palermo to the towns on the coast (Fig. 3).

The City of Palermo was excluded from the investigation, limiting the analysis to those in the territories outside the central area.[3] They are mostly complexes made up of detached or multi-family villas, mainly built in the last decade.

They vary in size, from complexes with a minimum of 3 houses to larger complexes where there are as many as 190 houses. There are varied facilities, with most complexes having swimming pools and sports facilities, and in a few cases, even

---

[3]The research was conducted through surveys, satellite observations and consultation of the portal of real estate ads or Internet sites of the same complex. The results are obviously partial, but together they return a significant starting point from which to deepen an almost unexplored and often overlooked topic.

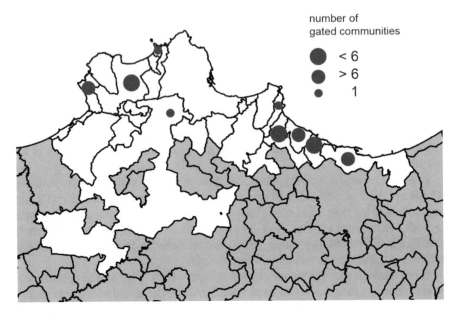

Fig. 3 Number of gated communities in the Metropolitan Area of Palermo (Image by Annalisa Giampino)

other types of services such as schools and nurseries. These are autonomous enclaves where the gated communities have surveillance and protection systems such as gates and video cameras, and even more structured forms of supervision, such as private security teams or guards at the gate. The largest number of these communities are found in the municipalities of Carini and Trabia.

It is plausible to think that the concentration in these areas is due to the proximity to the City of Palermo, and a transformation of seasonal complexes into permanent communities in more recent years. To support the hypothesis about the nature of gated communities in the complexes, there are the elements of protection and management of the communal parts and internal regulations. Compared to the possibilities offered by these residential complexes, self-exclusion in the low-density areas of the Metropolitan Area of Palermo finds justification in the lack of services in these areas, that are without equipment and infrastructures. In fact, the municipalities adjacent to Palermo have absorbed the city's excess population without a decentralisation policy for facilities or investment in the infrastructure and transport systems. Spatial self-segregation and self-organisation of services and infrastructures have become a necessity in this setting, while also being a desirable prospect (Ciulla 2011). We are therefore witnessing a progressive process of privatisation of space that is carried out via informal self-construction and via formal parcelling processes that were previously linked to seasonal complexes later transformed into permanent places of residence.

The development of new shopping malls is a recent phenomenon and, at the same time, shows the post-metropolitan transition that the area in question is undergoing. 8 shopping malls have been built in the Metropolitan Area of Palermo in recent years, with a total surface area of 675,900 sq.m (Giampino et al. 2014). The creation of these shopping centres, in relation to their complex implementation phase, have been publicised as the opportunity to bring public welfare to the *non-city* and, by no coincidence, represent one of the most important modes of development in recent years in areas outside the dense city of the Metropolitan Area of Palermo.

The actual transformations have been very different to the planned public and demagogically publicised objectives. A perfect example is the Poseidon Shopping Mall in Carini, a commercial centre of 32,851 sq.m that spreads over a surface area of 234,000 sq.m, built in one of the most problematic low-density areas of the metropolitan area.

The shopping mall, owned by Errichten srl, completed in 2010, was authorised by Carini Municipal Council Resolution 78 of 20 June 2010, in amendment to the *PRG* that allocated the area to tourist-hotel constructions (Fig. 4).

Without entering into the subject of the promptness of the procedure (although the primordial slowness that is characteristic of urban-planning-administrative practices in Sicily is well known), even in this case the construction of the shopping mall was promoted as being an opportunity for relaunching the local economy and at the same time as an economic opportunity to complete some essential public works for the area that had been waiting for funding for years.

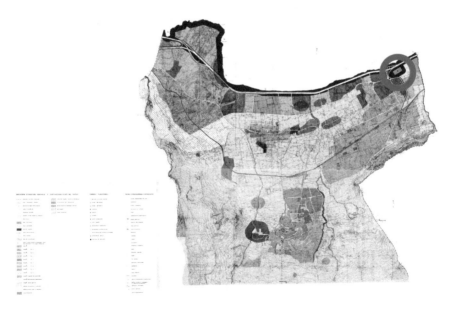

Fig. 4 Identification of the area of Poseidon Shopping Mall in the *PRG* of Carini (*Source Comune di Carini, PRG,* 1983)

In fact, one of the agreements stipulated to permit the amendment was the completion—by Errichten—of the stretch of road 7 of the regional road 5, with the relative bridge over the river Ciachea, which partly lies in the towns of Carini and Capaci.

A public work of primary importance, as can be seen in the Carini and Capaci Municipal Council resolutions, to decongest a heavily built-up area where the traffic from the new shopping mall would have made matters even worse. By the nature of the site it stands on, where there are various types of restrictions, nine years after the shopping mall opened, it is necessary to acknowledge the fact that the work has not been carried out, except for a section next to the car parks and serving the shopping mall.

In a territorial context where the public bodies find it difficult to provide adequate solutions, due to the progressive reduction in upgrading funds, the private economic operator has entered the scene, with financial possibilities, demanding ever-growing volumetric incentives that take place as described above.

The extent of the urban phenomena has provided the town councils with an honour, in terms of managing essential public services, without providing support for supra-local policies, placing these bodies in a situation where it is impossible to act. Through collusion, inertia or incapacity, the administrations in the Metropolitan Area of Palermo have limited themselves to negotiating development rights and projects without an ideologically or politically guided vision, remaining indifferent and deaf to the requests from the inhabitants of this large metropolis.

2.2   On the Lack of Policies

The true urban transformation policies have taken place in Palermo, while regional and local institutions tried to understand what type of administrative (rather than institutional) reorganisation was necessary to address the urban transborder dynamic. Without a general territorial reorganisation logic, the smaller centres around the large urban centres, as we have seen, have been affected by processes that they were not able to govern due to lack of funds, and to the difficulty in adapting to a spatial phenomenon that requires a sharing of roles and responsibilities, in the realm of a cooperative and multi-scale model. The intense political and academic debate, developed since the 1970s to date, restores the critical factors of public action.

The problem of the Palermo metropolitan territory has not been addressed in terms of reviewing the territorial model used, but in terms of administrative and political structure. By looking over literature produced in a period of time from the end of the 1970s to the start of the 2000s, what seems to most emerge is the problem of perimetrisation of the area.

A matter that pertains to privileges and prerogatives of municipal bodies and hypothetical new supra-local entities, rather than to the theory of reconfiguring territorial and functional layouts of the areas. Implicit policies, that can be found in the inconclusive and rhetorical political debate about administrative reorganisation on the Sicilian metropolitan body, show signs of advanced liberalism (Harvey 2012) and of the decline in the influence of public decisions on the actual territorial transformation processes, creating a precise action strategy.

As shown above, politics has not responded to the transfer of population with an improvement in services, with urban facilities and infrastructures, feeding the private dimension of living and a neoliberal, dominant transfiguration of multicentrism. The geography of the centres, with prestigious facilities and services, has been replaced by commercial centrality, thoroughly changing socio-spatial relations in the area.

These are urban areas indifferent to the landscape-environmental qualities they possessed, sold and distorted in the name of market laws and govern performance, welcoming commercial centrality and the *loisir* that the dense city did not want to/could not host.

The distance between the intention of public action and transformation has created a huge spatial lack of equality between dense city and the external areas. Areas born to welcome residential and commercial development that the city has expelled (without this dynamic being accompanied by an overall territorial reorganisation plan) and that force the people living in these areas into a tiring daily mobility. The institutions must inevitably build the new geography of public action on these topics, replacing the traditional urban categories.

3   Conclusions

The territories outside the inner cities of the new metropolitan areas of the South of Italy, as suggested by the metropolitan context we have analysed, show more contradictory, conflictual but also more interesting aspects than the contemporary urban phenomenon.

So far, the disciplinary view of these places has identified them as non-cities, considering it a difficult mater to find urban traces outside the traditional spatial and epistemological boundaries of the city (Brenner and Schmid 2015). The Palermo context, however, provides interpretative images that place the areas outside the dense centres on a much different plane to that of the anti-city.

In our investigation, the living dimension, intended not only as dwelling place, but also in the broader sense of socio-spatial organisation of ways of living in these areas, allowed us to uncover the connection of tangible and intangible relations at the basis of these types of urbanisation that are produced in the modern post-metropolis.

In a post-metropolitan dynamic marked by global economic geography and local identity construction processes, the case of Palermo, although considered marginal and distant from what would be an orthodox view of a post-metropolis, shows how the increasing privatisation processes of space are dynamic and affect all territories, confirming Brenner's most recent hypotheses (2016) about planetary urbanisation.

However, what emerges most from the Palermo case is that the post-metropolitan territory project is finding it hard to find answers and a place in the technical and political context.

The negotiations and difficulties caused by territorial bodies changing their minds are signs of public planning ability that has lost its predictive ability, meant as the capacity to interpret latent contradictions and micro-planning, taking refuge in an aseptic negotiation between collective interests and benefits, between economic players and the public, de facto excluding the production of public and planning (that is ambiguous, latent and conflictual), deriving from the interaction between space and inhabitants.

References

Alaimo G, Colajanni B, Pellitteri G (1996) I modi dell'abitare abusivo in Sicilia: analisi tipologica. Anvied, Palermo

Ardigò A (1967) La diffusione urbana. Ave, Rome

Astengo G (1990) Il metodo proposto. In: Astengo G, Nucci C (eds) IT.URB. 80. I dati della ricerca. Quaderni di Urbanistica Informazioni 2(8):4–7

Astengo G, Nucci C (eds) (1990) IT.URB. 80. I dati della ricerca. Quaderni di Urbanistica Informazioni 2(8)

Bianchetti C (2003) Abitare la città contemporanea. Skira, Milan

Brand R, Gaffikin F (2007) Collaborative planning in an uncollaborative world. Plan Theory 6(3):282–313

Brenner N (2016) Stato, spazio, urbanizzazione. Guerini e Associati, Milan

Brenner N, Schmid C (2015) Towards a new epistemology of the urban? City: Anal. Urban Trends, Cul, Theory Policy Act 19(2–3):151–182

Ciulla F (2011) Paradisi artificiali: Trasformazioni dello spazio simbolico e materiale nelle gated communities. Diritto e Questioni Pubbliche 11:601–629

Clementi A, Dematteis G, Palermo P (eds) (1996) Le forme del territorio italiano. Laterza, Bari

Costantino D (2008) Periferie metropolitane e forme insediative a Palermo. Planum: J Urbanism 17:1–12

De Carlo G (1962) La nuova dimensione della città. ILSES, Stresa

De Leo D, Lo Piccolo F (2015) Planning in the face of conflict in un-democratic contexts: lessons from two sicilian municipalities. In: Gualini E, Allegra M, Morais Mourato J (eds) Conflict in the city: contested urban spaces and local democracy. Jovis Verlag GmbH, Berlin, pp 80–93

Di Leo P (1997) Area metropolitana di Palermo. Città e Territorio. Bollettino del Dipartimento della Città e Territorio dell'Università di Palermo 3:72–79

Giampino A (2015) Gated communities a latitudini meridiane. In: Atti della XVIII Conferenza Siu, Italia 45–45. Radici, condizioni, prospettive, 11–13 June 2015, Venice. Planum Publisher, Rome-Milan, pp 879–885

Giampino A, Lotta F, Picone M, Schilleci F (2017) Sulle tracce della post-metropoli: l'area metropolitana di Palermo. In: Balducci A, Fedeli V, Curci F (eds) Oltre la metropoli. L'urbanizzazione regionale in Italia. Guerini e Associati, Milan, pp 193–221

Giampino A, Picone M, Todaro V (2014) Postmetropoli in contesti al 'margine'. Planum 2(29):1308–1316

Harvey D (2012) Rebel cities: from the rights to the city to the urban revolution. Verso, London

Hirschiman OH (1982) Shifting involvements: private interest and public action. Princeton University Press, Princeton

Indovina F (2005) Governare la città con l'urbanistica. Maggioli, Bologna

Indovina F, Matassoni F, Savino M, Sernini M, Torres M, Vettoretto L (eds) (1990) La città diffusa. Daest-IUAV, Venezia

Lo Piccolo F (2008) Il principio di cittadinanza attiva nella sua mutabilità interpretativa ed applicativa nell'ambito dei processi e degli strumenti di pianificazione. In: Lo Piccolo F, Pinzello I (eds) Cittadini e cittadinanza. Prospettive, ruolo e opportunità di Agenda 21 locale in ambito urbano. Palumbo, Palermo

Magnaghi A (ed) (1990) Il territorio dell'abitare: lo sviluppo locale come alternativa strategica. FrancoAngeli, Milan

McKenzie E (1994) Privatopia: homeowner associations and the rise of residential private government. Yale University, New Haven

Scott AJ (2008) Social economy of the metropolis: cognitive-cultural capitalism and the global resurgence of cities. Oxford University Press, Oxford

Secchi B (1994) Resoconto di una ricerca. Urbanistica 103:25–30

Secchi B (2013) La città dei ricchi e la città dei poveri. Laterza, Bari

Soja EW (2000) Postmetropolis: critical studies of cities and regions. Blackwell, Malden, MA

Soja EW (2011) Regional urbanization and the end of the metropolis era. In: Bridge G, Watson S (eds) New companion to the city. Wiley-Blackwell, Chichester, pp 679–689

Housing Affordability for Urban Regions

Giulia Bonafede and Grazia Napoli

Abstract Urban regions are recognised as driving forces of the global economy as well as the main sources of social inequality. In recent decades, particularly, housing access has become a serious problem, not only for the most disadvantaged population, but also for the middle class, as a result of economic crises and despite of prices decline in the housing market. In urban regions with high population density, some social groups face problems of housing affordability that depend not only on market prices but also on income availability. The contribution proposes a methodology for *income-threshold* assessment through a combination between the *ratio income* and the *residual income* approaches, which is applied to two different Sicilian urban regions: the urban region of Palermo (UR PA) and the urban region of Syracuse and Ragusa (UR SR-RG). A focus on low-income levels in high residential tension municipalities helps to highlight the distribution of the housing access problem. The analysis can provide a cognitive support for developing strategies and planning tools or for implementing actions that are oriented to address the housing issue.

1 Urban Regions and Housing Tensions

According to Soja (2000) the post-metropolis is the spatial result, albeit transient, of a new transformation of the city characterised by a new, high level of social and spatial fragmentation. This phenomenon, which was already characteristic of American and European geographic contexts, leads to a progressive convergence of demographic density in peripheral and central geographic areas, according to the American geographer. The polarisation of urban space, typical of the Fordist city, has in fact made way to an unstructured, dispersed pattern that traditional analytical

G. Bonafede (✉) · G. Napoli
University of Palermo, Palermo, Italy
e-mail: giulia.bonafede@unipa.it

G. Napoli
e-mail: grazia.napoli@unipa.it

F. Lo Piccolo et al. (eds.), *Urban Regionalisation Processes*,
UNIPA Springer Series,
https://doi.org/10.1007/978-3-030-64469-7_11

205

urban concentration models opposed to peripheral areas cannot interpret, requiring new analytical categories, such as urbanised regions or regional urbanisation, in an era that has been defined as that of "urban regionalisation" or "regionalised urbanisation" or "planetary urbanisation".

In this framework, the concept of post-metropolis is more a challenge to reflect on what is now intended with the term city (Balducci et al. 2017). The phenomena of fragmentation and dispersion, in fact, do not just involve large cities, but have also recorded a change in the urban structure of smaller towns too and in marginal geographic contexts, giving rise to urbanised regions (Bonafede et al. 2015) with no distinction between urban and rural, with complex and contradictory inter-scalar relationships.

Soja (2011) and Scott (2011) highlight the fact that from the second millennium onwards, the city-regions are not just the main drivers for global economy but are also the main generating sources of social inequalities and injustices, creating new issues about citizens' rights (Lefebvre 1970) and their controversial management.

In the large urban regions, the fracture among high- and low-income social groups has always been a constant phenomenon that has now been exacerbated (Scott 2011) and although the physical form of the city is continuously transforming, the right to the city in political terms remains in the era of globalisation, manifesting itself through the numerous urban movements that combat the dominant neo-liberal system (Harvey 2012). In particular, access to housing in recent decades has become a serious problem, not just for the most disadvantaged population bracket, increased by the migrant flow, but also for the middle class due to the economic crisis and in spite of the fact that property market prices have dropped. On the other hand, widespread urbanisation does not always correspond to demographic development, while growing land consumption is often combined with an increase in empty rooms.

This situation has been confirmed by the illegal occupations of public and private abandoned buildings and by the housing protest movements that still claim suitable policies for those who have the right to public residential building, but also for inhabitants who do not come under the required parameters. These parameters, that determine poverty income, or that define a boundary of this condition, vary depending on geographical areas, the size of the urban centres and the composition of the family (number of members, age, health problems).

In particular, *ISTAT* defines poverty by distinguishing urban areas in the various geographical areas of Italy depending on the demographic size (bigger or smaller than 50,000 inhabitants), while since 2003, the Inter-ministerial Committee for Economic Programming (*Comitato Interministeriale per la Programmazione Economica, CIPE*) has approved a list of municipalities in the regions of Italy that are defined as having High Housing Tension (*Alta Tensione Abitativa, ATA*) based on demographic growth.

Although vertical hierarchical relations can be seen in urban regions, between the various social classes in cities, or horizontal, synergic, complementary or mixed relations (Beguin 1979; Dematteis 1985; Scott 2011), it is doubtless that in these areas, overall housing density is high and the various social groups that receive low incomes must address *housing affordability* problems. Even if the location choices of residency are not solely based on the housing price/transport *trade-off* with the restriction of income and are influenced by other factors (cultural, aesthetic, ethical, identity, affections, etc.), the housing price is however recognised as a proxy of the location quality and technological and architectural characteristics of housing. Inhabitants' choices of location are therefore conditioned by the availability of income that can be allocated to the house and by market prices (for house purchase or rental). We therefore propose a *threshold income* evaluation methodology, using the combination of *ratio income* and the *residual income approach* that is applied to two territorial contexts: the former Metropolitan Area of Palermo (as defined by RL 9/1986) and a multi-nuclear urban region in the South-Eastern area of Sicily (Figs. 1 and 2).

A *focus* on income levels in municipalities that have already been defined as high housing tension, that fall within the two areas of study, will contribute to highlighting the housing access problem.

Fig. 1 View of Palermo from the *Palazzo Reale* (https://it.wikipedia.org/wiki/File:Vista_Pal ermo_dal_Palazzo_dei_Normanni5.jpg)

Fig. 2 Overview of Syracuse, South-Eastern Sicily (https://commons.wikimedia.org/wiki/File:Ort igia_dall%27alto.jpg)

2 The Concept of *Housing Affordability*

Different definitions of the concept of *housing affordability* have been offered: according to Burke (2004) it is the household's capacity to meet housing costs and basic costs of living; Hancock (1993) said that a rent is affordable when, after it has been paid, the consumer can have an acceptable standard of both housing and non-housing consumption; Hulchanski (1995) recognises that a household has a *housing affordability* problem when it pays more than a certain percentage of its income to live in adequate housing.

Housing affordability therefore addresses the housing issue from the point of view of demand that can also be supported via the provision of public subsidy for rents (Bonafede and Napoli 2015), but must be aided by the verification of housing quality standards (Fig. 3).

The fact that a household manages to purchase a house on the property market does not mean that the minimum housing standards are met (e.g. the house could be overcrowded or located in run-down buildings) and that its location in the outer area of the urban regions is avoided. In all cases, *housing affordability* is always a ratio between people and housing price:

- At the same price, it varies in correspondence to the household type, the income range and the minimum housing standards considered to be acceptable (Stone 2006);
- At the same income, it is affected by property market fluctuations. In fact, the capital gains (or capital loss) phases of the real estate (Rizzo 1999, 2002) increase

Fig. 3 Central area in Monreale, North-Western Sicily (Italy) (https://commons.wikimedia.org/wiki/File:Monreale_(40733451884).jpg)

(or reduce) the housing affordability problems (Napoli 2016) also affecting housing policies and town planning tools.

There are various approaches for measuring *housing affordability* (Stone et al. 2011), but the most common are the *ratio income approach* and the *residual income approach*.

The *ratio income approach* is an internationally diffused indicator, that expresses household ability/problem to pay for housing (HIA/CBA 2003; NAR 2005; Brookings Institution 2006; OMI 2017) in relation to a maximum *housing cost/income* ratio that is considered to be acceptable, which, once exceeded, means the household will not be able to meet the other needs. The value of this percentage is currently 30%.

This approach is criticised because the limit value is not obtained through statistical models but is a *rule-of-thumb*, and is also kept constant for any type of household or consumption model (Baer 1976).

The *residual income approach*, on the other hand, estimates how much a household can afford to spend for the house only after incurring costs for minimum needs.

In this approach, the indicator, that is the residual income, is not obtained from a ratio, but from a difference (Thalmann 2003; Stone et al. 2011).

3 Methodology

The proposed methodology introduces the concept of *threshold income (R_S)* as an indicator of *housing affordability* on a territorial scale, in order to contribute to the analyses supporting the setting of goals and implementation measures of housing policies.

The methodology foresees that the R_S is assessed according to territorial distribution of income and housing prices: the income bracket of taxpayers and the housing prices in central and semi-central areas are collected in each municipality in the area. Each resulting R_S therefore expresses a specific *housing affordability* and can be defined as the minimum income that allows a given household to purchase a house in a given location at a given price.

This study proposes to use the *threshold income R_S_combined* as an indicator, obtained by combining the approaches of the *ratio income* and *residual income* as indicated in the formula (1). The $R\_S\_residual$ is calculated first, to check that the income of a household I_j allows both housing costs and basic subsistence needs costs to be covered. The $R\_S\_residual$ obtained is then subjected to verification of the *ratio income*, checking that the housing cost is lower than a certain percentage of the household income. The differences between household incomes and the $R\_S\_residual$ are the income gaps that prevent a certain type of household from accessing the real estate market in a given area to be filled with public subsidies.

$$\begin{aligned} I_j - R\_S\_residual_{jmz} > 0 \\ R\_S\_residual_{jmz} > R\_S\_ratio_{jmz} \end{aligned} \quad \Rightarrow \text{housing affordability} \qquad (1)$$

The $R\_S\_residual$ is the minimum income that allows a given household to purchase a house at a given price in a given city and area, maintaining a certain amount of income fixed so that it covers basic household needs and is evaluated according to the formulas (2) and (3) specifying the minimum income that corresponds to the poverty threshold and almost poverty for any type of household, and minimum costs for meeting basic subsistence needs. The R_S_ratio is calculated according to the formula (4)

$$R\_S\_residual_{jmz} = P\&I_{mz} + NHE_{min-J} \qquad (2)$$

$$P\&I_{mz} = f(i, T, P_{mz}, LTV) \qquad (3)$$

$$R\_S\_ratio_{jmz} = \frac{P\&I_{mz}}{I_j * I_{ratio}} \qquad (4)$$

Where: $P\&I$ = mortgage instalment; NHE_{min-j} = minimum non-housing expenditure of the household type j; I_j = type j household annual income; I_{ratio} = affordable income ratio; i = annual mortgage interest rate; T = loan term; Pmz = housing price in area z of town m; LTV = loan to value.

$$Gap\_ratio_{jmz} = I_j - R\_S\_ratio_{jmz} \quad \begin{matrix} > 0 \text{ housing affordability} \\ < 0 \text{ housing affordability problem} \end{matrix} \quad (5)$$

According to formula (5), when the difference between the household incomes I_j and the R_S_ratio is lower (or greater) than zero, there are (there are no) *housing affordability* problems in a certain municipality or area.

To summarise the characteristics of the local real estate market, the threshold income has been estimated in correspondence with the maximum price of the central area *B1* and the minimum price of the semi-central area *C1* for the *OMI* type of "civil housing".

The parameters applied in the evaluation are contained in Table 1 and Fig. 4. The method was tested for a typical family of two adults (18–59 years) and a child (4–10 years). Obviously if the method is applied to families of different number and

Table 1 Parameters for calculation of R_S_ratio

Parameters	Unit	Value
Housing size	sq.m	70
Down payment	%	20
Loan to value (LTV)	%	80
Loan term	years	20
Annual interest rate (mortgage)	%	2.54
Monthly interest rate (mortgage)	%	0.21167
Mortgage instalments (P&I)	No.	240
Ratio income	%	30

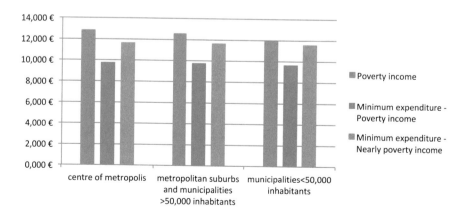

Fig. 4 Poverty income and minimum costs for a typical family in 2015 (Image by Grazia Napoli; data taken from *ISTAT*)

Fig. 5 Overview of Modica, South-Eastern Sicily (Italy) (Image by Giuseppe Abbate)

composition, the results will be different and the *housing affordability* problems will tend to increase or decrease (Fig. 5).

4 Case Study: The Two Urban Regions

The case study included two urban regions: the former metropolitan area of Palermo in North-Western Sicily and the multi-centre urban region in South-Eastern Sicily where the main cities are Syracuse and Ragusa (Fig. 6). The urban region of Palermo (UR PA) is part of the province of the same name (now a metropolitan city) and in 2014 had 1,074,495 inhabitants that reside in a surface area of 1,396 sq.km. The territorial system of Syracuse and Ragusa (UR SR-RG) includes the entire province of Ragusa and a part of the province of Syracuse, with, in 2014, a number of inhabitants totalling 551,749 units in a surface area of 2,600 sq.km, almost twice that of the other area included in the study (Fig. 6).

The two areas have a very different territorial structure: in the former case, the minor urban centres gravitate around the main city (Palermo), where 63% of the total area population is concentrated and where the executive departments and most advanced services are located, thus defining a mainly hierarchical and highly polarised territorial system (Giampino et al. 2017); in the latter case, the inhabitants of the main city (Syracuse) account for just 22.1% of the total area population and show that this is a multi-centre urban region made up of medium and small urban centres that form a *network* where horizontal relations prevail (Lo Piccolo et al. 2017).

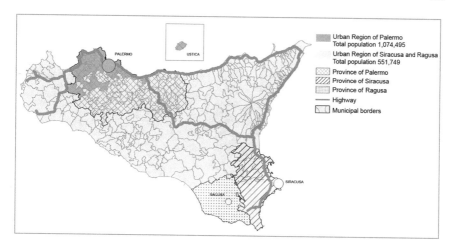

Fig. 6 The two urban regions (Image by Giulia Bonafede, data taken from *Comuni-Italiani.it*)

Highways serve both urban regions; nevertheless, in the South-Eastern area this kind of road infrastructure does not reach the city of Ragusa.

In addition, the two areas register high building densities on the coastal strips, which are compromised by an unbalanced touristic development with respect to internal areas. They also include numerous municipalities with marked population density and public housing lack for the low-income households, in spite of the high percentage of unused housing stock that in the Ragusa province is equal to 42% (Bonafede 2018).

4.1 The Urban Region of Palermo

The urban region of Palermo (Fig. 7) is formed by 27 municipalities, of which only eight come under the first circle surrounding with the administrative centre. Of these, Ficarazzi and Villabate form an urban *continuum* with Palermo in the plain of the southern coastal strip; Isola delle Femmine and Torretta extend over the northwest coastal hills; the municipalities of Belmonte Mezzagno, Altofonte, Monreale and Misilmeri extend over the inland hills. The nineteen remaining municipalities form the second or third ring and extend over the coastal and inland hills to the South and North-West.

In the area, there is an overall density of 769.7 inhabitants/sq.km (Fig. 8). The municipality with the highest population density is Villabate; followed by Palermo, Ficarazzi, Isola delle Femmine, Bagheria, Capaci and Balestrate. Of these municipalities, however, it must be pointed out that apart from Palermo, that has a surface area of about 159 sq.km, and Bagheria (Fig. 26), that measures almost 30 sq.km, the other municipalities extend over territories with very small surface areas, ranging from 3.5

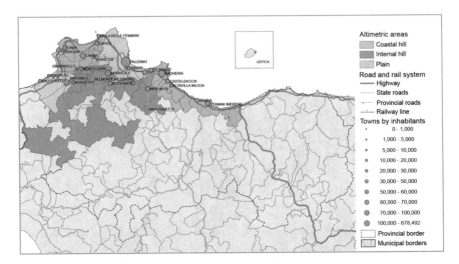

Fig. 7 Road and rail system, municipality by size and altimetric areas in the urban region of Palermo (Image by Giulia Bonafede, data taken from *Comuni-Italiani.it*)

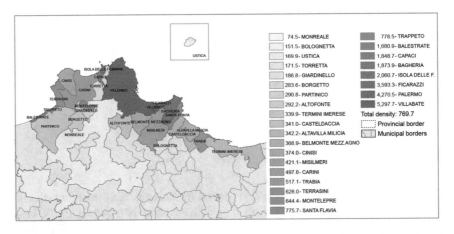

Fig. 8 Population density in 2014 (inhabitants/sq.km) by municipality in the urban region of Palermo (Image by Giulia Bonafede, data taken from *Comuni-Italiani.it*)

sq.km (Isola delle Femmine) to 6.1 sq.km (Capaci). The municipality with the lowest density is Monreale (74.5 inhabitants/sq.km), in the inland hills, that extends over a territory that is 30% larger than that of Palermo (Fig. 8).

In the urban region of Palermo, in the years 2002–2014, the percentage variation of the population (Fig. 9) shows that after a long period of decline, the Sicilian regional capital now only loses 1% of inhabitants, while Termini Imerese loses 2%, due to industrial closures.

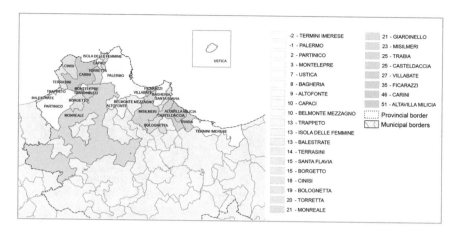

Fig. 9 Percentage variation (2002–2014) of population by municipality in the urban region of Palermo (Image by Giulia Bonafede, data taken from *Comuni-Italiani.it*)

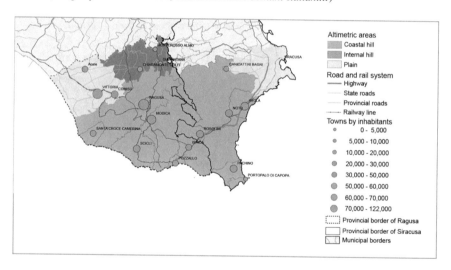

Fig. 10 Road and rail system, municipality by size and altimetric area in the Urban Region of Syracuse and Ragusa (Image by Giulia Bonafede, data taken from *Comuni-Italiani.it*)

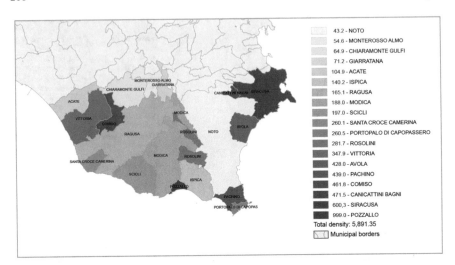

Fig. 11 Population density in 2014 (inhabitants/sq.km) by municipality in the Urban Region of Syracuse and Ragusa (Image by Giulia Bonafede, data taken from *Comuni-Italiani.it*)

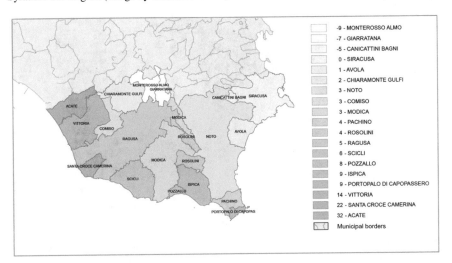

Fig. 12 Percentage variation (2002–2014) of population by municipality in the Urban Region of Syracuse and Ragusa (Image by Giulia Bonafede, data taken from *Comuni-Italiani.it*)

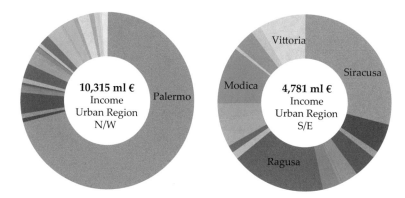

Fig. 13 Income produced in 2015 in the municipalities of the North-Western and South-Eastern urban regions (Image by Grazia Napoli, data taken from *Comuni-Italiani.it*)

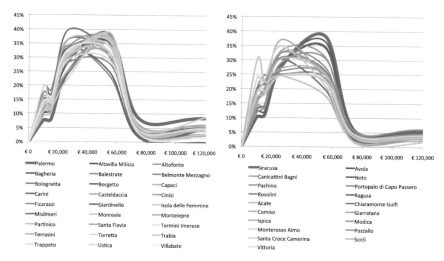

Fig. 14 Percentages by annual income brackets declared in 2015 by municipality. North-Western (left) and South-Eastern (right) urban regions (Image by Napoli 2017, data taken from *Comuni-Italiani.it*)

Fig. 15 Overview of Scicli, South-Eastern Sicily (Italy) (Image by Giuseppe Abbate)

Fig. 16 Ragusa, one of the two cities of the South-Eastern region of Sicily (https://commons.wik imedia.org/wiki/File:Ragusa_2008_IMG_1458.jpg)

Fig. 17 Overview of Isola delle Femmine with Palermo in the background (https://commons.wik imedia.org/wiki/File:Terra_mia,_L%27isola_delle_Femmine.jpg)

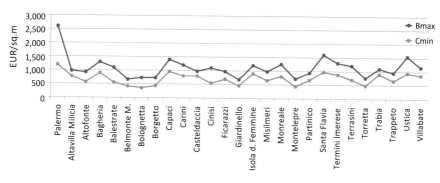

Fig. 18 Housing prices in EUR/sq.m by area and by municipality in the North-Western urban region (Image by Grazia Napoli, data taken from OMI)

All the other municipalities record an increase that in some cases is rather significant, such as Carini (46%) and Altavilla Milicia (51%).

In particular, the municipalities of Partinico, Montelepre, Ustica, Bagheria, Alto-fonte, Capaci and Belmonte Mezzagno show increases from 2 to 10%. The municipal-ities of Trappeto, Isola delle Femmine, Balestrate, Terrasini, Santa Flavia, Borgetto, Cinisi, Bolognetta, Torretta, Monreale, Giardinello and Misilmeri record increases

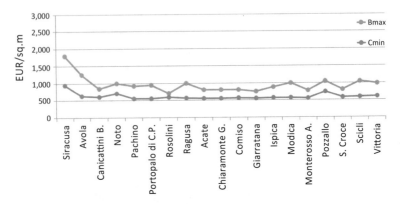

Fig. 19 Housing prices in EUR/sq.m by area and by municipality in the South-Eastern urban region (Image by Grazia Napoli; data taken from *OMI*)

Fig. 20 Housing in the central area of Modica (South-Eastern Sicily) (Image by Giuseppe Abbate)

between 13 and 23%. Variations ranging from 25 to 35% are found in the municipalities of Trabia, Casteldaccia, Villabate and Ficarazzi where the coastal anthropic pressure has been significant for some time (Fig. 9).

This means that the surrounding municipalities not only absorb the weak outflow from the city of Palermo (4,409 inhabitants) that is not compensated by the inflow of migrants, but over time also record an overall 4.7% increase in population, corresponding in absolute terms to 48,715 units.

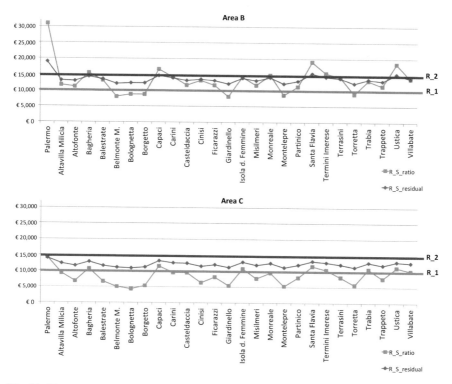

Fig. 21 Threshold incomes and incomes of the groups R_1 and R_2 in the areas B and C in the municipalities belonging to the North-Western urban region (Image and data by Grazia Napoli)

In all cases, since 2003 (*CIPE* deliberation 87) some municipalities that fall into the study area have been defined as *ATA* and have been included in a list[1] allowing special agreements to be stipulated between the state and municipalities to respond to critical situations.

In 2016, the Conference of Regions and Autonomous Provinces proposed a review of the thresholds for identifying *ATA* municipalities foreseeing to integrate the previous criterion based solely on demographic growth with an indicator of housing problem that takes into account the thorough territorial changes of the last ten years.

The variables that should contribute to defining this problem are the eviction/households ratio, the percentage of households in rented houses, the percentage of foreigners out of the total population and the percentage incidence of household poverty (see *Regioni.it* 2884 of 18 February 2016).

Although the definition criteria have been subject to review, the list published in 2004 has not yet been updated and in particular among the 12 *ATA* municipalities that

---

[1]The list of all *ATA* municipalities by Italian region is published in the Official Journal 40 of 18 February 2004.

Fig. 22 Central area in Modica, South-Eastern Sicily (Italy) (Image by Giuseppe Abbate)

fall into the study area, some urban centres with significant demographic increases, such as Altavilla Milicia, Carini and Casteldaccia have not been included.

4.2 The Urban Region of Syracuse and Ragusa

The South-Eastern urban region (Fig. 10) includes 19 municipalities, nine of which border directly with Ragusa while only three border with Syracuse. Most of the territory is made up of coastal hills. Only the municipalities of Acate, Vittoria and Comiso, in the province of Ragusa, as well as Syracuse and Canicattini Bagni in the province of Syracuse, are located on flat land. The only municipalities in the inland hills of the included Ragusa area are Chiaramonte Gulfi, Giarratana and Monterosso Almo.

Population density in the urban region of Syracuse and Ragusa is 213 inhabitants/sq.km (Fig. 11). The municipality with the highest recorded population density is the port town of Pozzallo (999 inhabitants/sq.km) with a rather small territorial surface area; then it is followed by the city of Syracuse (600 inhabitants/sq.km), the outlying small town of Canicattini Bagni (471 inhabitants/sq.km) and, in the opposite direction, the municipality of Comiso (462 inhabitants/sq.km) on the Ragusa plain.

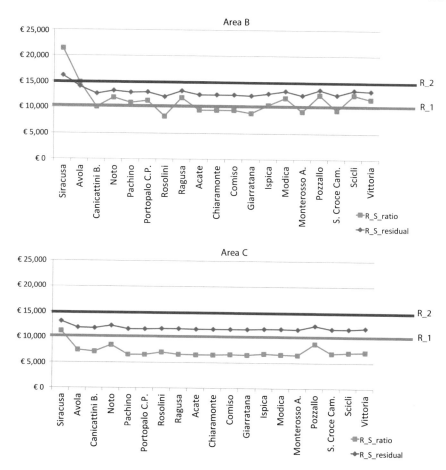

Fig. 23 Threshold incomes and incomes of the groups R_1 and R_2 in areas B and C in the municipalities belonging to the South-Eastern urban region (Image and data by Grazia Napoli)

Also the municipalities of Pachino, Avola and Vittoria record fairly high population density (ranging from 439 to 348 inhabitants/sq.km) in relation to the area in question. In the larger area of Noto, population density is the lowest of all. Low densities, under the threshold of 100 inhabitants/sq.km, follow in Monterosso Almo, Chiaramonte Gulfi and Giarratana, where declines in population numbers have been recorded (Figs. 11 and 12).

Modica and Ragusa with their large territories, record rather moderate housing densities, that together with Acate, Ispica and Scicli range between 100 and 200 inhabitants/sq.km. Santa Croce Camerina, Portopalo di Capo Passero and Rosolini have between 200 and 300 inhabitants/sq.km.

A 4.6% increase in population has also been recorded in the South-Eastern system (Fig. 12). The towns that lose inhabitants are the innermost municipalities of

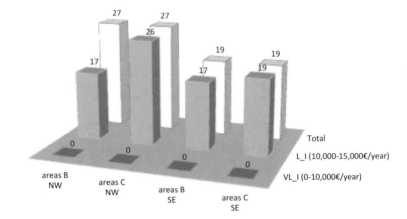

Fig. 24 *R_S_combined*. Frequency of housing affordability by income bracket and by area in the North-Western and South-Eastern urban regions (Image and data by Grazia Napoli)

Fig. 25 Housing in the central area of Scicli (South-Eastern Sicily) (Image by Giuseppe Abbate)

Monterosso Alma, Giarratana and Canicattini Bagni, while the population trend in the city of Syracuse is quite stable.

Almost all municipalities record population increases of 1–9%. Ragusa is halfway with an increase of 5%. Larger increases have been recorded in the municipalities of Vittoria (14%), Santa Croce Camerina (22%) and most of all in Acate (32%), although

Fig. 26 Bagheria in the urban region of Palermo (https://it.wikipedia.org/wiki/File:Bagheria.jpg# metadata)

the latter two urban centres are not included in the list of seven *ATA* municipalities in the South-Eastern area.

4.3   *Territorial Distribution of Income*

Wealth analysis expressed in terms of annual income declared in 2015 (*ISTAT* data), confirm polarisation of the Palermo urban region. In fact, the city of Palermo has the concentration of the more profitable economic activities that produce 70.9% of the total income.

On the other hand, in the South-Eastern urban region, the income produced in the city of Syracuse is just 28.8% and is indicative of a more balanced distribution of wealth among the municipalities (Fig. 13).

In the urban region of Palermo (Fig. 14 left), the income brackets 15,001–55,000 €/year are the most frequent, even if some municipalities have high percentages of very low and low incomes (0–15,000 €/year). The highest income brackets are found in the lowest percentages. In the South-Eastern urban region (Fig. 14 right) the low-income brackets in many municipalities account for up to 30%; in others, the percentages are higher in the 15,001–26,000 €/year, or 26,001–55,000 €/year brackets (e.g. Syracuse, Ragusa and Noto) (Fig. 15).

4.4 Average Incomes and a Focus on ATA Municipalities

Income levels, together with housing market prices, are the parameters that are used in determining the threshold income as part of the methodology already described in section three.

The Ministry of Economy and Finance (*Ministero dell'Economia e delle Finanze, MEF*) and the website *Comuni-Italiani.it* publish taxable income data for the purpose of additional municipal income tax and for a complex class system that are not always the same, showing the relative frequency of taxpayers and the corresponding total in Euros.

For research purposes, therefore, a choice was made to distinguish the average income by overall taxpayers (RD), the average income by population (RP) and the average income of taxpayers up to 10,000 €/year (RD*) as under this threshold, according to ISTAT 2014 calculations on absolute poverty, in urban regions of the South of Italy, there are certainly families with an income of poverty for an average family of three members (two adults and one child).

Another two easily indicators are considered that can be deduced from data published for each municipality, i.e. the percentage of taxpayers compared to the populations and the percentage of taxpayers up to 10,000 €/year (D*) compared to the total number of taxpayers (D).

The data from the two urban regions (27 municipalities for the North-Western area and 19 municipalities for the South-Eastern area) have been aggregated and summarised in Table 2, while data from municipalities with *ATA* falling within the two urban regions have been analysed individually in Tables 3 and 5, distinguishing the income bracket up to 10,000 €/year in Tables 4 and 6 respectively.

In 2014, in Sicily, the average income of taxpayers was 15,959 €/year with a 55.6% of the population declaration percentage, which is lower than the Italian percentage, while the average income compared to the whole population is below the poverty threshold and below the Italian average (Tables 2 and 3).

Table 2 Average incomes by taxpayers and % of Declarations compared to the Population by geographical areas in 2014 (Processed by Giulia Bonafede; data taken from *MEF* and *Comuni-Italiani.it*)

Geographical areas	RD €/year			RD %		D/P %	
	2002	2014	Variation	2002	2014	Variation
UR PA	11,178	14,846	33	50.7	49.8	−0.9
UR SR-RG	10,208	13,523	32	61.3	59.9	−1.4
Sicily	12,311	15,959	30	57.7	55.6	−2.1
Italy	15,640	20,299	30	69.4	66.1	−3.3

RD average income of taxpayers, *D* number of taxpayers, *P* population

Table 3 *ATA* Municipalities, declared incomes by taxpayer and average incomes within the urban region of Palermo in 2014 (Processed by Giulia Bonafede; data taken from *MEF* and *Comuni-Italiani.it*)

Town/region	Taxpayers (D)	Population (P)	D/P %	Total amount €/year	RD €/year	RP €/year
Altofonte	5,436	10,307	52.7	84,758,352	15,592	8,223
Bagheria	26,307	55,615	47.3	392,685,184	14,927	7,061
Capaci	5,443	11.3	48.1	83,702,203	15,378	7,398
Ficarazzi	5,649	12.79	24.2	85,208,104	15,084	6,661
Isola delle Femmine	3,643	7.29	59.9	62,677,432	17,205	8,592
Monreale	19,113	39,410	48.5	300,561,045	15,725	7,627
Misilmeri	13,424	29,143	46.1	187,557,994	13,972	6,436
Montelepre	3,273	6,373	51.4	47,039,454	14,372	7,381
Palermo	354,789	678,492	52.3	7,259,266,441	20,461	10,699
Trabia	5,417	10,579	51.2	76,530,514	14,128	7,234
Torretta	2,124	4,358	48.7	29,564,645	13,919	6,784
Villabate	8,826	20,290	43.5	112,657,910	12,764	5,552
Total area ATA	453,444	885,968	51.2	8,722,209,278	19,235	9,845
Sicily	2,832,310	5,092,080	55.6	45,202,089,768	15,959	8,877
Italy	40,205,353	60,795,612	66.1	816,119,791,979	20,299	13,424

D number of taxpayers, *P* population, *RD* taxpayers' average income, *RP* population's average income

From the analyses carried out in the two urban regions (UR PA and UR SR-RG), it was found that the average income by taxpayers (RD) in the period of time 2002–2014 is lower than the Sicilian one, although there was an increase in percentage variation (33 and 32%) that is higher than the Italian and the Sicilian percentage variations (Table 2) and in spite of the fact that the percentage of declarations in relation to the population (D/P) decreases over the considered time range.

In fact, while the average income of taxpayers does not naturally align with the values of larger geographical areas (Sicily and Italy), even if it increases, the reduction in percentage of declarations is less consistent than the Sicilian and the Italian ones, which in all cases confirm a generally negative trend (Table 2). It was also found that the percentage of income declarations in the urban region of Syracuse and Ragusa is higher than the Palermo one and the Sicilian one (Tables 3 and 4).

In particular, in the 12 *ATA* municipalities of the urban region of Palermo (Table 3), the average income of taxpayers (RD) is below the Sicilian average; only Palermo and Isola delle Femmine (Fig. 17) record higher incomes.

Palermo in particular aligns with the average income of taxpayers in Italy but not with the average income of the Italian population.

Table 4 *ATA* Municipalities and Incomes declared up to 10,000 €/year within urban region of Palermo in 2014 (processed by Giulia Bonafede; data taken from *MEF* and *Comuni-Italiani.it*)

Town	Taxpayers (D*)	Population (P)	D*/P %	Total amount €/year	RD* €/year	D*/D %
Altofonte	2,209	10,307	21.4	10,050,648	4,550	40.6
Bagheria	11,678	55,615	21.0	52,409,354	4,488	44.4
Capaci	2,257	11,314	19.9	10,623,772	4,707	41.5
Ficarazzi	2,283	12,792	17.8	10,729,208	4,700	40.4
Isola delle Femmine	1,408	7,295	19.3	6,478,319	4,601	38.6
Monreale	8,040	39,410	20.4	38,230,787	4,755	42.1
Misilmeri	6,034	29,143	20.7	28,012,308	4,642	44.9
Montelepre	1,381	6,373	21.7	6,502,568	4,709	42.2
Palermo	123,860	678,492	18.3	585,023,735	4,723	34.9
Trabia	2,425	10,579	22.9	10,444,542	4,307	44.8
Torretta	994	4,358	22.8	4,295,638	4,322	46.8
Villabate	3,837	20,290	18.9	17,871,409	4,658	43.5
Total area ATA	166,406	885,968	18.8	780,672,288	4,691	36.7

*D** number of taxpayers up to 10,000 €/year, *P* population, *D*/P* incidence of taxpayers on population, *RD** average income of taxpayers, *D*/D* incidence of taxpayers D* on the rest of taxpayers D

Villabate, that is the urban centre with the highest population density, also has the lowest average income, and the lowest percentage of taxpayers, followed by Ficarazzi.

Low incomes were also recorded in the municipalities of Misilmeri, Trabia and Torretta.

In all 12 *ATA* municipalities, the population percentage that declares an income below or equal to 10,000 €/year (D*) compared to the rest of taxpayers (D) is over 40%, apart from Isola delle Femmine and Palermo (Table 4) that have percentages (D*/D) of 38.6 and 34.9 respectively. If the data for Isola delle Femmine (Fig. 17) is not very significant, considering the small number of inhabitants, the Sicilian capital's data is also corroborated by the high percentage of taxpayers compared to the population (see Table 3). Higher percentages of income declarations below or equal to 10,000 €/year are recorded in Trabia and Torretta. Overall in the 12 *ATA* municipalities, for those that declare they come under this income bracket, there is a recorded annual average income (RD*) of 4,691 €/year that corresponds to less than 400 € per month (Table 4).

In the seven *ATA* municipalities of the South-Eastern urban region, there is a higher average income than the rest of Sicily only in the provincial capitals (Table 5). On the other hand, Vittoria and Canicattini Bagni have the lowest average income. Overall in the *ATA* municipalities in this area, the percentage of taxpayers compared to the population is higher than the urban region of Palermo and almost all the municipalities

Table 5 *ATA* Municipalities, declared incomes by taxpayer and average incomes within Urban Region of Syracuse and Ragusa in 2014 (Processed by Giulia Bonafede; data taken from *MEF* and *Comuni-Italiani.it*)

Town/Region	Taxpayers (D)	Population (P)	D/P %	Total amount €/year	RD €/year	RP €/year
Avola	16,590	31,785	52.20	252,723,993	15,234	8,223
Canicattini B.	4,157	7,124	58.40	57,836,628	13,913	8,119
Modica	32,715	54,651	59.90	492,146,618	15,043	9,005
Noto	2,944	23,834	54.30	199,829,116	15,438	8,384
Ragusa	49,095	73,030	67.20	821,147,497	16,726	11,244
Syracuse	71,588	122,503	58.40	1,367,818,681	19,107	11,166
Vittoria	36,318	63,092	57.60	406,948,566	11,205	6,450
Tot *ATA* area	213,407	376,019	56.75	3,598,451,099	16,862	9,570
Sicily	2,832,310	5,092,080	55.60	45,202,089,768	15,959	8,877
Italy	40,205,353	60,795,612	66.10	816,119,791,979	20,299	13,424

D number of taxpayers, *P* population, *RD* taxpayers' average income, *RP* population's average income

align with the Sicilian data, apart from Avola and Noto. Ragusa (Fig. 16) in particular is the most virtuous, with a higher percentage of taxpayers than the Italian data, while Syracuse records the highest average income (Table 5).

With regard to the income bracket lower than or equal to 10,000 €/year (Table 6), the highest percentage of taxpayers (D*/P) is found in Canicattini Bagni and Vittoria.

Table 6 *ATA* Municipalities and Declared incomes up to 10,000 €/year within the Urban Region of Syracuse and Ragusa in 2014 (Processed by Giulia Bonafede; data taken from *MEF* and *Comuni-Italiani.it*)

Town	Taxpayers (D*)	Population (P)	D*/P %	Total amount €/year	RD* €/year	D*/D %
Avola	7,069	31,785	22.2	33,255,600	4,704	42.6
Canicattini B.	1,914	7,124	26.9	8,608,355	4,498	46.0
Modica	13,722	54,651	25.1	67,890,462	4,948	41.9
Noto	5,527	23,834	23.2	24,408,982	4,416	42.7
Ragusa	19,532	73,030	26.7	87,562,288	4,483	39.8
Syracuse	25,321	122,503	20.7	109,736,245	4,334	35.4
Vittoria	21,331	63,092	33.8	100,342,507	4,704	58.7
Tot *ATA* area	94,416	376,019	25.1	431,804,439	4,573	44.2

*D** number of taxpayers up to 10,000 €/year, *P* population, *D*/P* incidence of taxpayers on population, *RD** average income of taxpayers, *D*/D* incidence of taxpayers D* on the rest of taxpayers D

In Vittoria, in particular, almost 59% of those who declare income come under the income bracket between 0 and 10,000 €/year, but the lowest average incomes are found in Noto (4,416 €/year) and Syracuse (4,334 €/year).

Overall in the seven *ATA* municipalities in the South-Eastern urban region, the average income of this particular income bracket is equal to 4,573 € per annum and is lower than the corresponding data recorded in the urban region of Palermo (see Table 4).

It can therefore be inferred that the South-Eastern urban region data are on average more realistic considering the fact that the percentage of taxpayers is higher.

In the period 2002–2014, the income levels in the two urban regions showed an upward trend (Table 2) but higher percentages of taxpayers in the lower income bracket were recorded, making the gap between social groups more marked. The comparison between the two urban regions shows that the income level is higher overall in the Palermo area, while the percentage of declarations is higher in the South-Eastern area.

Also, when referring to the *ATA* municipalities, the 0–10,000 €/year income bracket is more common in the South-Eastern area (Tables 4 and 6), with an average income (4,573 €/year) that is lower than the one (4,691 €/year) in the Palermo area.

4.5 The Threshold Income in the Two Urban Regions

In order to estimate the threshold income, housing prices were collected in the *OMI B1* and *C1* areas for all municipalities in the North-Western and South-Eastern urban regions, and price of *B5* and *C11* areas in the city of Palermo.

The highly polarised territorial structure in the North-Western urban region is reflected in the large difference between prices in the central area of the main city and the other municipalities. In fact, the average of *Bmax* prices in the *hinterland* is 41% of the *Bmax* price in Palermo.

Higher prices in the *hinterland* are found in the municipalities with larger populations or that are closer to Palermo (e.g. Termini Imerese, Capaci), or in the ones with a thriving tourist economy (e.g. Ustica) (Fig. 18).

Housing prices in municipalities in the South-Eastern area are more homogeneous and reflect the existence of a multi-centre region.

The average of *Bmax* prices in the *hinterland* reaches 50% of the *Bmax* price in Syracuse (Fig. 19).

The *R_S_combined income* was calculated for the *B1* and *C1* areas in all municipalities in the two urban regions, applying the formulas 1–4 and the parameters from Table 1 and Fig. 4.

To check for the presence/absence of *housing affordability* problems in the most vulnerable social groups, the *R_S_combined* was compared with the lower *ISTAT* income brackets: the *R_1* group with an income of 0–10,000 €/year and the *R_2* group with an income of 10,001–15,000 €/year. It follows that 38.3 and 14% of taxpayers' income in the North-Western urban region and 27 and 10.3% in the South-Eastern region, i.e. almost over or half the total, belong to the groups R_1 and R_2 respectively. The analysis of any *housing affordability* problems for these two social groups is therefore significant for defining social policies or preparing *social housing* and public housing projects.

If the verification of conditions for formula (1) is applied to the group R_1, combining the R_S_ratio and $R\_S\_residual$ ratios evaluated for the different areas, it is found that this group cannot afford to live in any area of any municipality in the North-Western urban region (Fig. 20). On the other hand, *housing affordability* is rather satisfactory for the group R_2 that can purchase a house in almost all the semi-central areas C and central areas B of many municipalities in the region, except, in particular, the municipalities of Palermo, Bagheria, Capaci, Santa Flavia and Ustica, that are the regional capital and the towns with high tourist presences respectively (Fig. 21).

The income of the group R_1 is below the *threshold income* to access all the areas of all the municipalities even in the South-Eastern urban region (Fig. 22). *Housing affordability* is verified for taxpayers in the group R_2 for all the areas of all the municipalities except for the central area of Syracuse and Avola (Fig. 23).

These results produce significant implications from the point of view of territorial analysis; in fact it shows, on one hand, that households belonging to the group R_1 (38.3 and 27% of taxpayers respectively in the North-Western and South-Eastern urban regions) cannot afford to buy a house and need subsidies to rent even in the outskirts of small municipalities. The group R_2, on the other hand, in spite of the low-income bracket, has a good *housing affordability* in both urban regions, even if excluded from the capital city and the most important towns (Figs. 23, 24, and 25).

5   Conclusions

As highlighted by Scott (2011) for the American geographical context, the two Sicilian urban regions also show slight increases in income levels, while inequalities have increased more rapidly with high percentages of population crowding the lower income bracket.

While the convergence of demographic growth between outlying and central areas, as suggested by Soja (2000), is controversial, it appears to be clear that housing access for large portions of the low-income population is difficult in both urban regions, whether it has network, multi-centre characteristics or whether they develop in a highly polarised form. In fact, both areas have peripheral territories in decline, as they suffer from the historical process of transferring anthropic pressure from inland hill areas to the coastal plains.

In particular, in the *ATA* municipalities, the highest percentages of population with low incomes show that in all cases, integrated housing policies are urgently needed, such as to increase subsidies for rent or subsidised rental contracts (to reintroduce private empty housing onto the market), to create public housing (by renovating abandoned public buildings) and to plan family aid programmes.

This study has analysed the *housing affordability* problems in the urban regions using the *threshold income,* with the aim of defining the spatial distribution of both urban areas that are affordable/non-affordable to the low-income classes, and the corresponding income gaps that separate households' affordability from the housing market prices.

The results of the *combined-income approach,* applied to the case study of two Sicilian urban regions, show that the reduction in real estate market prices has reduced *housing affordability* problems, making house buying more affordable for low-income social classes. Social groups with a very low income, on the other hand, need to be supported not just financially, but also via planning that can modify the urban settlement.

In the North-Western urban region, for example, polarisation of the territory is behind the large differences in housing prices and social polarisation. The use of *threshold income* to identify the critical areas (due to low household income brackets or high housing prices) may be useful to support housing policies, public plans and actions for preventing low-income households from leaving the main cities or creation of ghettos in degraded areas on the edge of the urban regions.

References

Baer WC (1976) The evolution of housing indicators and housing standards: some lessons for the future. Public Policy 24(3):361–393

Balducci A, Fedeli V, Curci F (2017) Italia post-Metropoli? In: Balducci A, Fedeli V, Curci F (eds) Oltre la metropoli. L'urbanizzazione regionale in Italia. Guerini e Associati, Milan, pp 9–38

Beguin H (1979) Urban hierarchy and the rank-size distribution. Geogr Anal 11:149–164

Bonafede G (2018) Patrimonio abitativo. In: La Greca P, Vinci I (eds) Sicilia. Rapporto sul territorio 2018. Urbanistica Dossier 16:35–39

Bonafede G, Lo Piccolo F, Todaro V (2015) Migratory transnational flows as analytical tool for planning in the post-metropolitan transition of Italian urban regions. In: Gospodini A (ed) Proceedings of the international conference on changing cities II: spatial, design, landscape socio economic dimensions. Porto Heli, Peloponnese, Greece, 22–26 June 2015, Grafima Publ, Thessaloniki, pp 2513–2520

Bonafede G, Napoli G (2015) Palermo Multiculturale tra gentrification e crisi del mercato immobiliare nel centro storico. Archivio di Studi Urbani e Regionali 113:123–150

Brookings Institution (2006) The Affordability Index: a new tool for measuring the true affordability of a housing choice. https://www.brookings.edu/wp-content/uploads/2016/06/20060127_affindex.pdf. Accessed 7 July 2017

Burke T (2004) Measuring housing affordability. Australian Housing and Urban Research Institute, Melbourne

Dematteis G (1985) Contro-urbanizzazione e strutture urbane reticolari. In: Bianchi G, Magnani I (eds) Sviluppo multiregionale: teorie, metodi, problem. FrancoAngeli, Milan, pp 121–132

Giampino A, Lotta F, Picone M, Schilleci F (2017) Sulle tracce della post metropoli: l'area metropolitana di Palermo. In: Balducci A, Fedeli V, Curci F (eds) Oltre la metropoli. L'urbanizzazione regionale in Italia. Guerini e Associati, Milan, pp 193–221

Hancock KE (1993) 'Can pay? Won't pay?' or economic principles of affordability. Urban Stud 30(1):127–145

Harvey D (2012) Rebel cities: From the right to the city to the urban revolution. Verso, London and NewYork

HIA/CBA (2003) Housing report: a quarterly review of housing affordability. https://www.com mbank.com.au/about-us/news/media-releases/2003/Commonwealth-Bank-HIA-Housing-Rep ort.pdf. Accessed 7 July 2017

Hulchanski JD (1995) The concept of housing affordability: six contemporary uses of the housing expenditure-to-income. Hous Stud 10(4):471–491

ISTAT (2015) La povertà in Italia. https://www.istat.it/it/archivio/202338. Accessed 7 July 2017

Lefebvre H (1970) Il diritto alla città. Marsilio, Padua

Lo Piccolo F, Picone M, Todaro V (2017) South-eastern Sicily: a counterfactual post-metropolis. In: Balducci A, Fedeli V, Curci F (eds) Post-metropolitan territories: looking for a new urbanity. Routledge, New York, pp 183–204

Napoli G (2016) The economic sustainability of residential location and social housing: an application in Palermo City. Aestimum Jan: 257–277. http://www.fupress.net/index.php/ceset/article/view/17896/16729. Accessed 7 July 2017

Napoli G (2017) Housing affordability in metropolitan areas: the application of a combination of the ratio income and residual income approaches to two case studies in Sicily, Italy. Buildings 7(95):1–19

NAR-National Association of Realtors (2005) Housing affordability index. http://www.realtor.org. Accessed 28 Feb 2016

OMI (2017) Rapporto immobiliare 2017. Il settore residenziale. http://www.agenziaentrate.gov. it/wps/content/Nsilib/Nsi/Documentazione/omi/Pubblicazioni/Rapporti+immobiliari+residenzi ali/. Accessed 7 July 2017

Rizzo F (1999) Valori e Valutazioni. La scienza dell'economia o l'economia della scienza. FrancoAngeli, Milan

Rizzo F (2002) Dalla rivoluzione keynesiana alla nuova economia. Dis-equilibrio, tras-informazione e coefficiente di capitalizzazione. FrancoAngeli, Milan

Scott AJ (2011) Città e regioni nel nuovo capitalismo. L'economia sociale delle metropoli. il Mulino, Bologna

Soja EW (2000) Postmetropolis: critical studies of cities and regions. Blackwell, Malden, MA

Soja EW (2011) Beyond postmetropolis. Urban Geogr 32(4):451–469

Stone ME (2006) A housing affordability standard for the UK. Hous Stud 21(4):453–476

Stone ME, Burke T, Ralston L (2011) The residual income approach to housing affordability: the theory and the practice. http://works.bepress.com/michael_stone/7. Accessed 7 July 2017

Thalmann P (2003) House 'poor' or simply 'poor'? J Hous Econ 12(4):291–317

Printed in the United States
by Baker & Taylor Publisher Services